# RECONSTRUCTING URBAN REGIME THEORY
## Regulating Urban Politics in a Global Economy

EDITED BY
# Mickey Lauria

SAGE Publications
*International Educational and Professional Publisher*
Thousand Oaks  London  New Delhi

*For information address:*

SAGE Publications, Inc.
2455 Teller Road
Thousand Oaks, California 91320
E-mail: order@sagepub.com

SAGE Publications Ltd.
6 Bonhill Street
London EC2A 4PU
United Kingdom

SAGE Publications India Pvt. Ltd.
M-32 Market
Greater Kailash I
New Delhi 110 048 India

Printed in the United States of America

*Library of Congress Cataloging-in-Publication Data*

Main entry under title:

Reconstructing urban regime theory: regulating urban politics in a global economy/
  edited by Mickey Lauria.
      p.  cm.
    Includes bibliographical references and index.
    ISBN 0-7619-0150-7 (cloth: acid-free paper).—ISBN 0-7619-0151-5
  (pbk.: acid-free paper)
      1. Municipal government.  2. Municipal government—United States.
    3. Local government.  4. Local government—United States.
    I. Lauria, Mickey.
JS78.R43   1996
320.8'5—dc20                                                        96-25361


97  98  99  00  01  02  03  10  9  8  7  6  5  4  3  2  1

*Acquiring Editor:*        Catherine Rossbach
*Production Editor:*       Diana E. Axelsen
*Production Assistant:*    Karen Wiley
*Typesetter/Designer:*     Marion S. Warren
*Indexer:*                 Jean Casalegno
*Print Buyer:*             Anna Chin

# Contents

# Preface

▬▬▬▬▬▬

The purpose of this project is to bring together scholars who have begun to write about the need for and advantages of conceptualizing urban politics within higher-level abstractions. Urban regime theory is the chosen perspective on urban politics precisely because it dispenses with the stalled debates between elite hegemony and pluralist interest group politics, between economic determinism and political machination, and between external and/or structural determinants and local and/or social construction. Rather, urban regime theory provides a robust conceptual framework that views these, on the one hand, as false dualisms and, on the other hand, as theoretically driven historical or empirical questions. The new urban politics literature in general, and urban regime theory in particular, has been criticized for relying solely on middle-range concepts and thus failing to interpret their contextual meaning via higher-level abstractions. In other words, their middle-level concepts are not theoretically well defined for the structure of capital and the role of the state.

The attraction of regulation theory is that it offers a way of linking changes in the economy to those in politics at a high level of abstraction. Here, the state, and local politics, are part of the mode of regulation within a particular regime of accumulation. But regulation theory has a tendency to underestimate the causal efficacy of local agents and institutions. Stronger conceptual linkages must be established between local agents and institutions, regime transformation, and the restructuring of urban space. Thus, the integration of the two approaches—the project of this book—appeared quite attractive.

In conceptualizing this edited volume, I decided to try to build an integrated collective project. Thus, the project began at the conceptualization stage with e-mail discussions about urban political economy, urban politics, and planning theory. Quite early, it became clear that two distinct projects were emerging: (a) reintegrating political economy and planning theory and (b) reconceptualizing urban regime theory within the context of a regulationist approach. The former project was developed as a paper session at the 37th Annual Conference of the Association of Collegiate Schools of Planning (October 1995) and a symposium "Planning Theory and Political Economy" for the journal *Planning Theory* (No. 14, 1996). The latter project was developed as two paper sessions at the 26th Annual Meeting of the Urban Affairs Association (March 1996) and has birthed this edited book.

As organizer/editor, I collected individual queries, comments, and positions and bounced them back to a collection of interested scholars through an e-mail distribution list. After a few months of collecting and distributing comments, comments on comments, and so on, I requested abstracts of potential chapters for an edited volume. Unfortunately, not all who participated in these early discussions could find time in their schedules to produce chapters for the book. Regardless, I think it is appropriate to recognize the early contributions of Pierre Clavel, Susan Fainstein, Paul Knox, David Perry, Robert Whelan, and David Wilson.

After developing a book prospectus from the chapter abstracts and securing a book contract, I distributed the prospectus and a guide to chapter contributors with a selective bibliography to ensure a coherent focus. Although the chapters did not receive blind reviews and thus cannot be considered refereed, each chapter was reviewed by at least three other contributing authors[1]. The reviewers sent their comments directly to the contributing authors with copies to me. I should note that the contributor reviews were professional and provided invaluable guidance for the authors. Subsequently, as editor, I reviewed each chapter and the peer reviews and developed a suggested plan of revision for each contributor. Here I should note that the contributors responded openly and effectively to the suggested plan of revision. The goal of the contributor review process was not only to ensure quality control and provide me with more intellectual force from which to persuade contributing authors of the value of the suggested revisions but also to allow each reviewing contributor to read relevant/tangential chapters in the volume before revising their own contributions. The effect is a more coherent volume.

# Note

1. Realizing that this was an unusual and time-consuming process, I secured token monetary compensation for each contributing author ($100) as part of their chapter contract from Sage.

# 1

# Introduction

## *Reconstructing*
## *Urban Regime Theory*

### MICKEY LAURIA

Urban regime theory appears to have gained the dominant position in the literature on local politics precisely because it dispenses with the stalled debates between elite hegemony and pluralist interest group politics, between economic determinism and political machination and between external or structural determinants and local or social construction. Rather, urban regime theory provides a robust conceptual framework that views these as false dualisms and, alternatively, as theoretically driven historical or empirical questions. Urban growth coalitions are viewed as only one of the political coalitions that may arise in cities and need to be hegemonic for an extended period to be considered a corporate controlled regime. At the same time, entrepreneurialism is viewed as only one possible leadership approach that local politicians and government bureaucracies can pursue. Thus, urban regime theory asks how and under what

AUTHOR'S NOTE: I have borrowed liberally from a section of a paper that has appeared in *Planning Theory* (Lauria & Whelan, 1995). I want to thank the editor and publisher for permission to reproduce pages 12 through 15 here.

conditions do different types of governing coalitions emerge, consolidate, and become hegemonic or devolve and transform.

The recent transference of urban regime theory to contexts outside the United States and its use in comparative cross-national research also attests to its dominant position in urban political scholarship. Not surprisingly, these uses have further exposed some of the theory's shortcomings. I argue that it is necessary to reconceptualize the middle-level abstractions of the urban regime by interpreting it through the lens of the higher-level abstractions within regulationist theory (see, e.g., Fainstein, 1995; Feldman, 1995). Regulation theory is useful in directing one's attention when one is developing middle-range specifications. From its higher-level gaze, we know to evaluate the local structure of capital and the various fractions of capital within the governing coalition, to look for emerging institutional arrangements tied to consensus seeking and social regulation, and to evaluate the external connections of local politicians and capitalists. In this chapter, I will indicate the potential benefits of cross-fertilization between regulation theory and urban regime theory. My goal is to suggest how one might go about reconstructing a more viable urban regime theory that maintains its robust characteristics.

## Regimes and Urban Politics

The original focus of urban regime theory was to abstract models of urban politics from historical epochs (that some) connected to changes in the structure of the world economic system. What is interesting about urban regime theory are not the typologies it inspires but, rather, the structural features, the axes, or the defining focus that scholars use to describe local politics and abstract to typologies. Fainstein and Fainstein (1983) argue that the character of urban regimes derives from two structural features of the political economy: first, that local government organizationally depends on property taxes for its fiscal solvency; thus, municipalities must maintain their revenues by enhancing existing property investments and attracting more such investment. At the same time, low-income populations are a drain on local revenues. Second, a basic feature of a capitalist economy is the private control of production. The population at large depends on private investment and profit for employment. Thus, the state must "facilitate accumulation in order to advance the material interests of its citizens" (p. 251). These structural features explain business dominance in local politics. Fainstein and Fainstein, in their analytic summary of the case studies on the political economy of urban redevelopment in the United States, describe three successive post-World War II urban regimes: *directive* (1950-1964), *concessionary* (1965-1974), and *conserving* (1975-?) regimes (pp. 159-168). In the directive regime, the governing coalition planned large-scale redevelopment that was directly sponsored by local government. The difference between this regime and the subsequent concessionary regime was not a change in business dominance but, rather, the addition of socially and politically forced concessions made to lower-class urban residents. The conserving regime is capital striking back: retracting concessions and conserving political and social control while maintaining fiscal stability in the tenuous national and world economy.

Regime succession, according to Fainstein and Fainstein (1983, pp. 276-279), depends on the character of political struggles (social movement pressures) and the extent of fiscal or economic constraint. The post-WWII economic expansion in the United States lasted until approximately 1973. The political pressures of the civil rights movement and the welfare rights movement forced concessions on local political regimes, but the post-WWII economic expansion allowed these concessions both fiscally and politically. Thus, the succession from a directive to a concessionary regime was politically and economically expedient. The end of the post-WWII economic boom forced an end to these concessions and spurred the development of the conserving regime.

Elkin argues that the structural features that define an urban regime stem from what he calls the "division of labor between market and state," specifically, the definition of the respective prerogatives of the "controllers of productive assets" (1987, pp. 18-35) (capital) and public authorities, the organization of public authority, and the external relations of public authorities to private controllers of productive assets. These structural features were politically constructed during the urban (governmental) reform movement of the late 19th and early 20th centuries (Elkin, 1987, p. 19). These features suggest three defining axes of urban politics: public and private growth alliances, electoral politics, and bureaucratic politics. According to Elkin (pp. 36-60), urban political economies are defined by the particular constellation of these axes—that is, the particular ways in which the various land interests and politicians ally, electoral coalition strategy, and the structure of bureaucratic service provision. He discusses three such constellations that vary temporally and geographically: *pluralist* (Northeast and Midwest, 1950s through early 1960s), *federalist* (Northeast and Midwest, mid-1960s through late 1970s), and *entrepreneurial* (post-WWII Southwest) political economies and their respective regimes. Some have since argued that this entrepreneurial form is no longer confined geographically (Harvey, 1989). Although the focus and, thus, the names are different, Elkin's first two regime types are essentially the same as Fainstein and Fainstein's. The two schemes depart in the third regime type, with Elkin's being more pertinent to the 1980s and early 1990s. Fainstein's subsequent scholarship (1990, 1994) seems to indicate concurrence with my assessment.

Responding to this periodicity, urban political scholars began to indicate that the cities they study do not demonstrate the expected constellation during the specified epoch. In other words, there is more geographical and temporal variation than the Fainsteins or Elkin had suggested. The various chapters in Stone and Sanders (1987) exemplify this variation (see also Clavel, 1986; Cummings, 1988; Jones & Bachelor, 1986; Logan & Swanstrom, 1990). This geographic variation was interpreted by some to suggest that the connections between urban regimes and structural changes in the world economy were extremely complex if not spurious (see Beauregard, 1989b; Smith & Feagin, 1987, for exceptions). Stone (1987, p. 16) does not suggest that these connections are spurious but, rather, he assumes them to be mediated by local political and economic actors. Stone (1987), in his analytic summary of this edited book, developed a new typology of urban regimes that used governing coalition structure and development outcomes as their defining feature: *corporate, progressive,* and *caretaker.*

In his second book on Atlanta, Stone (1989) fully explicated his version of regime politics. Through this source, one clearly sees the theoretical roots of Stone's approach (see pp. 3-12, 179-199). The basis and focus of Stone's approach mirrors Banfield's (1961) *Political Influence,*[1] albeit less formal and historically richer. If one recalls, Banfield's main problematic was that government and economic power were so fragmented in metropolitan areas (the pluralist assumption), it was a wonder anything was ever accomplished. So Banfield's goal was to uncover how this formal decentralization of power (again political and economic) was politically overcome (temporarily centralized) to achieve a specified end. Banfield used a microeconomic investment model of human behavior to explain how actors used their various forms of influence to develop political coalitions that either promoted or stymied large-scale development projects with the ultimate goal of accumulating more political power. His descriptive generalizations stressed the importance of organizational actors in development politics. Uncertainty played a major role in actors' evaluations of the costs and benefits of taking a political position, exerting influence, or maintaining informal semiautonomous partially centralized structures of influence.

Stone (1989) promotes a *social production model* that is based on a similar question: "How, in a world of limited and dispersed authority, do actors work together across institutional lines to produce a capacity to govern and to bring about publicly significant results?" (pp. 8-9). Stone's coalition building process is informal, built on the uncertainty (see p. 9) of the various forms of influence,[2] and his actors are also concerned with the cost/benefit analysis of maintaining those coalitions. According to Stone,

> Elite power in Atlanta is therefore constrained, not so much by the counter-vailing power of others outside the coalition as by the maintenance needs of the governing coalition itself. (p. 195)

Following Banfield, Stone (1989) focuses internally on local coalition building, theoretically neglecting the connections to the larger world economy (see also Harding, 1994, p. 376):

> Instead of these alternatives [elite domination, pluralism in the form of veto groups, electoral politics, or apolitical economic enhancement], I have suggested that the internal politics of coalition building best explains why various policy initiatives took the particular form that they did. (Stone, 1989, p. 178)

Regime formation and transition are based on the stochastic microeconomic investment calculations of individual actors involved in coalition building and maintenance. Connections to external (state and national) politics and the larger world economy are considered only insofar as they affect those individual actors' calculations. This exclusion began what some have called a volunteerist return in urban politics (witness Stone's, 1993, attempt to distinguish his work from pluralist theory, pp. 3-7). As indicated earlier, Stone does not deny these structural constraints; in fact, he assumes them, but by focusing solely on the local individual and organizational actors involved in coalition building, he risks slighting or neglecting to evaluate the respective effects of those constraints.[3] Consequently, the abstraction of theoretical insights becomes confined to behavioral microeconomic, and possibly pluralistic, explanations of the social production of cooperation and political coalition building (see also Stone, 1993; Stone, Orr, & Imbroscio, 1991). To avoid this tendency, we must revert back to a Fainsteinesque approach that focuses on the connections to external political economic relations. Regulation theory potentially offers a framework to do so.

## Regulating Urban Politics

Regulation theory emerged out of a critique of structural Marxist theory. To put it simply, these theorists posed the question (also often posed by non-Marxian political economists): If the logic of capitalist accumulation is so strewn with (strong) contradictions, how has it survived so long? Regulation theorists hypothesized that extraeconomic (in the narrow sense of the word) forces help respond to the economic contradictions in a fashion that leads to the long-term survival of the basic capitalist social relationships, albeit in new forms that contain new contradictions. These forces are said to regulate the regime of accumulation. Thus, regulationist theory is concerned with the regulation of the processes of capital accumulation within a particular capitalist mode of production. This regulation is not presupposed by the logic of the accumulation process itself. It depends on a series of social, cultural, and political supports that are only contingently codependent. Regulation theory attempts to construct a historically and geographically grounded account of capitalism's development.

Regulationists argue that capital accumulation is reproduced through the dynamic of the prevailing *regime of accumulation*. The regime of accumulation specifies the broad relationships between production, consumption, savings and investment, and the geographical extent and degree of autonomy of the capital circuits. But the particular forms of these relationships have changed over time and across space because of adjustments to inherent contradictions and resultant crises. The regulating mechanisms include social institutions, social relations in civil society, cultural norms, and the activities of the state apparatus. A particular crystallized combination is referred to as a *mode of social regulation*. Similarly to the typological tendency imbued in urban regime theory, some authors have devolved regulation processes into a typological mode of analysis with historical periods being classified by the nature of their respective regime and mode of regulation.

The attraction of regulation theory is that it offers a way of linking changes in the economy to those in politics at a high level of abstraction (see Chouinard, 1990; Florida & Feldman, 1988; Florida & Jonas, 1991; Goodwin, 1992; Goodwin, Duncan, & Halford, 1993; Jessop, 1990a, 1995b). Here, the state and local politics are part of the mode of regulation within a particular accumulation regime. According to Goodwin et al. (1993, p. 75), the state is part of the mode of regulation of a *fordist* regime of accumulation because the state helped stabilize the patterns of production and demand (via Keynesian economic management), consumption through providing appropriate forms of service provision (via the various components of the welfare state), fiscal policy, wage relations (via collective bargaining and corporatist politics), and productive infrastructure. This fordist mode of regulation and the state's role corresponds to the epoch that comprises Elkin's *pluralist* and *federalist* and Fainstein and Fainstein's (1983) *directive* and *concessionary* urban regimes. The recent economic restructuring and transition from a fordist to a postfordist accumulation regime have been used to explain the rise of an *entrepreneurial* urban regime (see Goodwin et al., 1993; Harvey, 1989; Jezierski, 1994; Krätke & Schmoll,

1991; Stoker, 1990) that facilitates privatization and the dismantling of collective services. Although regulation theorists argue that the transition from one mode of regulation to another will be uneven and spatially differentiated, this is explanation by caveat because of the differentiated structure of urban regimes.

The first solution is to look for helpful explanations at a lower level of abstraction (see following). For example, we know that economic restructuring occurs unevenly, so we can hypothesize that differing external economic constraints can begin to explain the differentiation in specific urban regime structures and policies (see Axford & Pinch, 1994, pp. 357-358; DiGaetano, 1989; DiGaetano & Klemanski, 1994; Hall & McIntyre Hall, 1993/1994; Leo, 1995a; Logan & Swanstrom, 1990; Whelan, Young, & Lauria, 1994). Second, we can hypothesize that the structure of capital and representation of capital fractions locally will affect the nature of the governing coalition, potential urban regimes and their urban development strategies (see Horan, 1991; Lauria, 1986, 1994a).

The authors in this volume find this typologic approach limiting, often leading to functionalist logic, and far too static to capture the historical dynamics of our changing social formation. These authors prefer to be less concerned with deciphering static modes of regulation and instead focus on understanding dynamic regulatory mechanisms.

The ontological perspective is that of a social constructionist—the structural logic of capitalism and society's institutional organization, social forms, political processes, and cultural norms are mutually constituting. Although the structural logic of capitalism, as manifested in regulatory processes, constrains and informs social practices, social practices also constrain, inform, and constitute the concrete forms of capitalist regulatory processes. Given this, a historical or geographical account of the development cannot be built on the structural logic of capitalism, but concrete research is necessary to inform or reconstruct our abstract understandings of the logic of capitalism. Thus, a description of regulatory mechanisms does not explain the emergence of particular social practices.

# Are Urban Regime and Regulation Theory Complementary and Compatible?

Regulation theory appears to offer a fruitful set of abstractions in which to embed urban regime analysis because they have complementary strengths and weaknesses. Although urban regime theory inadequately theorizes the connections between local agents (economic and political) and their wider institutional context, regulation theory underestimates the importance of local actors and organizations and thus cannot explain the concrete construction of regulatory mechanisms. Regulation theory focuses on extralocal political and economic influences whereas urban regime theory focuses on the machination of political practices. Both have an overarching concern with governance: regime theory with political coalitions and their capacity to govern and regulation theory with the governance of production and consumption systems.

This marriage does not portend a solution to all our theoretical and methodological problems, for both theoretical approaches inadequately conceptualize scale: Urban regime theory underestimates the value of extrametropolitan spatial scales, and regulation theory's abstractions ignore spatial variations in material and discursive practices. This suggests a more important problem: Urban regime theory tends to be tied to causal relations explained with rational choice theory, whereas regulation theory is more open to causality being based on more than individual rational actions. But regulation theory has not grappled with material and discursive practices, so it has yet to demonstrate how that broader theory of causality would operate. Thus, if urban regime theory is to provide the analysis of material and discursive practices, it must do so without resorting to purely rational behavioral processes. Finally, both undertheorize capitalism: Urban regime theory has no explicit theory short of the division between market and state, whereas regulation theory reduces capitalism's complexity to discrete transformations between homogeneous accumulation regimes ignoring how material practices constitute modes of regulation. Thus, much theoretical reconstruction must be done. What is certain is that the basis of that theoretical reconstruction must be empirical research focusing on the concrete social practices of urban politics in specific places and times. At the same time, that empirical research must be reflexively previsioned with careful abstractions that attempt to resolve the aforementioned deficiencies of the urban regime and regulation theories.

## Notes

1. This is not a completely pejorative attribution, for I view Banfield's book as one of the few classics in urban politics. It has stood the test of time and now has received the ultimate compliment of being retread as the foundation of a new school of thought (the new urban politics or urban regimes) almost 30 years later. Ironically, the subtitle of the book is *A New Theory of Urban Politics*.

2. See his use of Banfield's types of influence, reciprocity, and loyalty, to name a few, on page 180.

3. This is also a self-criticism. One main critical point a colleague made of a recent paper (Whelan, Young, & Lauria, 1994) was that we failed to discuss the role of the State of Louisiana in our analysis of regime transition in New Orleans. My response at the time was, "We cited your paper

[Miron, 1992] on the fiscal reform movement in Louisiana that does just that; that in this paper we can't do everything." In retrospect, that omission was an indicator of the larger neglect of the external political-economic forces involved.

# P A R T 1

## Conceptualizing the Regulation of Urban Regimes

I n Part I, the authors grapple with the currently understood deficiencies of the regulationist approach and urban regime theory, then attempt to resolve new problems inherent in a complementary synthesis.

Although rehearsing some cogent criticisms of the concept "mode of regulation," Goodwin and Painter argue that this does not fatally undermine the regulationist project. They argue for a methodological approach that focuses on the ebb and flow of regulatory processes and practices rather than on the construction of static or crystallized modes of regulation and stable regimes of accumulation. Their regulation theory, as a method, seeks to account for the spatial and temporal uneven reproduction of capitalist social relations. For Goodwin and Painter, social practices and regulatory processes are mutually constituted in space and time and thus are inherently uneven. This constructivist ontological position forces attention to the dialectical relationship between theory development and concrete research (retroduction). Their interest in regulation and concrete research has directed them toward an investigation of urban regimes in an attempt to understand local governmental actors' cooperation with private actors in managing and securing economic develop-

ment: what they, and others, call the shift from government to governance
so manifest in the United Kingdom. Goodwin and Painter argue for an
investigative frame that "joins a geographically sensitive regulation
theory based on social processes to a critical political sociology of the
local state and urban governance based on an investigation of the material
and discursive practices in which they are grounded." They leave open
the question of whether urban regime theory can contribute to this critical
sociology of the local state.

Feldman follows Goodwin and Painter's lead in his attempt to develop
a theory of spatial structures necessary for understanding urban regime
formation. He retheorizes regulation theory, considering the causal properties of space. He develops a new conceptualization, spatial structures
of regulation (SSR), to situate urban regimes with respect to six circuits
of capital that compose a particular SSR. He argues that an analysis of a
locality's SSR would help identify issues that are important to local
capital and help explain the nature of their involvement in local politics.
In addition, such analysis could help us better understand the effects of
local public policy. More important, Feldman argues that urban regime
theory can help us understand the local state's role in the social construction of the local SSR. Obviously, this requires that such urban regime
analysis be embedded within a regulationist approach.

Finally, Jessop provides us with the basis for a reoriented urban regime
research agenda. His interpretation draws on the complementarity of
Gramscian state theory and regulation theory to contextualize urban
regime analysis. From this basis, Jessop signals lessons to heed during
analysis of local economic governance. I will mention only three. He
asserts that one must understand how the local economy becomes constituted as an object of supralocal economic and extraeconomic regulation. Second is to evaluate the relationship between local economic
strategies and prevailing hegemonic projects. Third, he argues that it is
important to analyze how institutions and apparatuses are strategically
selective. Particular forms of economic and political systems privilege
some strategies over others, some forces over others, and so forth. Jessop
argues that the durability of urban regimes depends on the coherence and
economic feasibility of their strategies and on the strategic capacities
rooted in local institutional structures and organizations, that is, how the
urban regime is linked to the formation of a local hegemonic bloc and
its associated historical bloc.

C
H
A
P
T
E
R

# 2

# Concrete Research, Urban Regimes, and Regulation Theory

MARK GOODWIN
JOE PAINTER

I n this chapter we explore some of the epistemological and methodological implications that arise from any attempt to integrate urban regime theory and regulation theory. In particular, by examining the essentially methodological character of regulation theory, we examine the claim that regulation theory offers higher-level abstractions with which to complement the lower-level abstractions

AUTHORS' NOTE: The authors would like to thank the editor for his efforts in bringing this volume together and for his comments on this particular chapter. Bob Beauregard, Kevin Cox, Bob Jessop, and Christopher Leo also kindly made constructive comments on an initial draft. We must also thank the Economic and Social Research Council (ESRC), whose funding of project No. L311253011 Local Governance in the Transition From Fordism made these thoughts possible in the first place. A longer version of some of the arguments presented here can be found in an article titled "Local Governance and Concrete Research: Investigating the Uneven Development of Regulation," published in *Economy and Society,* Vol. 24, No. 3, pp. 334-356, 1995.

of regime theory. We contend that the implicit counter positioning of a more empirical regime theory with a more abstract regulation theory sets up a false dichotomy that ignores, or at best understates, the ways in which regulation theory can be used to set up and inform concrete research. Such concrete research can, in turn, draw on a reconstituted regime theory, a point that several chapters in this book develop in more detail.

Our investigations lead us to propose that regulationist research should be less concerned than hitherto with the identification of more or less coherent "modes" of regulation and more interested in the ongoing processes of regulation constituted through material and discursive social practices. A concern with such practices highlights both the significance of the concrete and the role of space and geography in regulation. We suggest that processes of regulation are constituted through a plethora of unevenly developed social practices—including, for instance, those of urban governance and urban regimes. This means that they can be investigated only by giving due weight to concrete research.

This is for two reasons. First, the processes that define abstract "modes of regulation" arise only in concrete social practice. All regulation is generated in temporally and spatially embedded social contexts. Epistemologically, therefore, the abstract depends on the concrete and *rational* abstractions (Sayer, 1984) that can exist only insofar as they are genuinely abstractions from (and, conversely, grounded in) concrete circumstances. Second, the inevitable and complex uneven development of regulation raises the possibility that spatial differentiation itself affects regulatory processes both quantitatively ("how much" regulation) and qualitatively ("what kind" of regulation). Investigating this possibility again requires concrete research because specifying elements of regulation in the abstract cannot establish how, and with what effects, regulation elements interact in particular geographical contexts. Nor, of course, does an abstraction establish how they vary over time. In what follows we will develop these arguments in more detail before examining how regulation theory and regime theory might be brought together in concrete research.

## Regulation Theory and Methodology

In trying to develop regulationist methodologies, we start with the arguments of Robert Boyer (1990)[1] and Bob Jessop (1990a) who have been particularly concerned to restate the "original methodological concerns of the pioneer regulation theorists" (Jessop, 1990a, p. 153). Many different approaches to studying the political economy have drawn on regulation theory, and there are several different regulationist "schools" (for a survey, see Jessop, 1990a). Boyer's own substantive work forms part of the so-called Parisian school, but it seems to us that his methodological discussion has a more general applicability.

Stressing methodology allows us to focus on regulation theory as a perspective and a form of analysis that is in principle distinct from the substantive claims made about particular structures and processes by individual regulationist writers. We want to explore regulation theory as an approach without necessarily endorsing all the historical accounts (about fordism as a mode of regulation, for example) developed in its name.

Virtually all explicitly regulationist work to date has been concerned with the regulation of the economy or, more precisely, with the regulation of the process of capital accumulation within the capitalist mode of production. Without regulation, it is argued, the process of capital accumulation would collapse through its own contradictions. If stable or expanded, accumulation occurs, it does so partly as a result of a process of regulation *that is not presupposed by the accumulation process itself.* According to regulation theory, continued accumulation depends on a series of social, cultural, and political supports that are only contingently copresent. Moreover, there is not just one possible pattern of regulation but, at least in principle, many alternative contingent combinations of "noneconomic" factors that might operate to support accumulation, with varying degrees of effectiveness.

Regulation theory lies within the broad tradition of Marxist historical materialism. It is concerned with the temporal and spatial variability of capitalism.[2] Marx's abstract accounts of the "laws of motion" of capital and the "necessary tendencies" of the accumulation process are fundamentally significant for political economy. But their very abstraction means that these accounts cannot, in themselves, provide an accurate account of the historical (and geographical) development of the capitalist mode of production. One can derive from Marx an account of certain necessary tendencies toward spatial and temporal uneven development as David Harvey (1982) has shown. The regulationist project is somewhat different, however. Though not denying the importance of *necessary* tendencies to uneven development, regulation theory focuses principally on the *non*necessary relations that determine which specific periods and places see what kinds of accumulation and economic growth. This oversimplifies somewhat and we will elaborate. For now, the significant point is that regulation theory aims to develop a historically and geographically grounded account of capitalism's development. In this sense, the regulation approach itself must be linked with higher-level abstractions.

According to Marxist historical materialism, although there is a necessary tendency to the expanded reproduction of the social relations of capitalism, there are also necessary counterprocesses in the form of crisis tendencies (of overaccumulation or underconsumption, for example). Regulation theory aims to explain the actual ebb and flow between these tendencies in historical time. It seeks to specify the different institutional and cultural circumstances that led, among other things, to the crisis of the 1930s and the sustained growth of the 1950s and 1960s. Spatially, regulation theory focuses on the uneven development of accumulation as a result, in part, of the spatially differentiated development of those institutional and cultural conditions.

Thus, the work of Marx allows us to identify Japan, the United States, and Brazil as "capitalist" countries. It also justifies labeling as capitalist both late 19th century capitalism and late 20th century capitalism. By contrast, a regulationist perspective can help explain how and why capitalism is different in each country and how and why capitalism in the late 19th century differs from capitalism today. Insofar as these differences are nonnecessary, rather than presupposed by the defining features of the abstract accumulation process, then the regulationist development of historical materialism (or something very like it) is required to explain them. Insofar as those differences are significant for the people involved or partly implicated in the historical development of the capitalist mode of production, or both, their explanation seems to us a desirable goal.

In the more economistic or vulgar uses of Marxism, theoretical argument often began with the accumulation process and its necessary tendencies and implications (and sometimes went no further). Insofar as wider social and political questions were raised, they tended to take the form of investigating the consequences of the accumulation process for the state, culture, social relations, or whatever. Although Marx himself was aware of the social embeddedness of capitalist relations and often emphasized the interplay between state, society, and economy, in the writings of many of his followers the implicit (and sometimes explicit) direction of causation was held to be from the economy to the forms of the state and civil society. Social life and political processes and institutions were taken as the *explanandum* (that which is to be explained), and the process of capital accumulation (or aspects thereof) was advanced as the *explanans* (that which explains).

There is a sense in which regulation theory reverses that "causal arrow." That is, the success or failure and stability or instability of the accumulation process is held to be, if not wholly determined, at least heavily influenced by institutional organization, social forms, political processes, and cultural norms. This perhaps overstates the case a little, and it is important to recognize that the process of capitalist accumulation is not compatible with every conceivable social, cultural, and political form. The capitalist accumulations process does coexist with a very wide variety of such forms, however, and according to regulationist methodology, temporal and spatial variations in the nature and combination of these nonnecessary processes are crucial in determining the course of capital accumulation at different times and in different places. Of course, we must add to this the necessary tendencies to spatial and temporal variation that Harvey argues are inherent in the accumulation process itself.

All this places regulation theory within the context of a longer tradition of Western Marxism (and to some extent contemporary post-Marxism). The Frankfurt School, Althusserian structuralism (with its concept of overdetermination), Marxist historians, Gramscian cultural studies, and contemporary discourse theory all emphasize the relatively or even wholly autonomous effectivity of noneconomic aspects of social life.

The idea that regulation theory simply reverses the direction of causality is an oversimplification, however. First, regulation theory does not deny that the process of capital accumulation has significant consequences for social life in capitalist societies. Second, as we will discuss in more detail, regulationist accounts of economic change work at a greater degree of concreteness and complexity than the classical Marxist discussion of accumulation. Although not denying the Marxist account of the logic of capital, regulation theory recognizes that account is highly abstract and contains contradictory tendencies and countertendencies. Regulation theory proposes that the character of the concrete manifestation of those abstract tendencies and countertendencies is by definition not predictable on the basis of the abstractions themselves but is determined only at a more concrete degree of analysis ("concretization") and once influences from other aspects of social life are considered ("complexification") (Jessop, 1990a, p. 165).

In this sense, therefore, regulation theory turns the progress of the accumulation process into its explanandum and changing and geographically varied political, cultural, and social forms into its explanans. By extension, therefore, regulation theory cannot in itself provide a complete explanation of the changing character of the state or culture or social relations because these are by definition to some extent part of its explanatory tools, not the objects of its explanations.[3]

As a method of analysis, then, regulation theory starts from the premise that the reproduction of capitalist social relations is not guaranteed by the abstract relations that are the defining features of the capitalist mode of production. Indeed, as we pointed out at the beginning of the chapter, these abstract relations can be realized only in concrete social practice. Hence, both crises in the reproduction process and in the phases of expanded reproduction (when these occur) are the products of more concrete institutional structures, political and social processes, and cultural discourses. Although the abstract features of capitalism as disclosed by Marx are transhistorical, all these more concrete forms and practices vary historically and geographically. The reproduction of capitalist relations in historical time and geographical space is thus a highly uneven process. Regulation theory as a method seeks to explain that uneven pattern. This requires not only a sensitivity to history and geography but also a clear understanding of what is being explained, the causal processes proposed as explanation and the degrees of complexity and concreteness used.

## Modes, Processes, and Sites of Regulation

### Mode of Regulation: A Redundant Concept?

In investigating the uneven pattern of capitalist reproduction, many regulationist accounts have adopted the concept of the mode of regulation. Although we are not seeking to abandon this terminology entirely, we do believe that it is problematic and needs to be used with caution. As we have noted, regulation theory explains the spatially and temporally uneven reproduction of capitalism with reference to the variable pattern of cultural and political institutions and practices. Conventionally, the term *mode of regulation* refers to a specific combination of these operating together in a mutually reinforcing way. Thus, the mode of regulation known as fordism for example, involved technical change in the process of production, organizational change in capital-labor relations, political implementation of Keynesian demand management policies and the emergence of a social norm of mass consumption for manual workers and their families. Most regulationists argue that these practices and developments interacted to form a "virtuous circle" of rising wages, productivity, and output underwritten by the welfare state.

We broadly accept this analysis, although we would stress the uneven spread of these arrangements, both within and between countries, and have pointed out elsewhere that the virtuous circle had some socially discriminatory effects (Bakshi, Goodwin, Painter, & Southern, 1995). There are, however, several difficulties with labeling the set of relations as a mode of regulation.

First, the term *mode* is often understood as implying a completed system, rather than one in the process of formation. Critics argue that the notion of modes of regulation overemphasizes the functionality, stability, and coherence of regulatory relations and underemphasizes change, conflict, and development during their period of operation. Second, by extension, the idea of contrasting modes succeeding one another places too much stress on sharp breaks and radical discontinuities in the development of capitalist societies. A crude account of one stable and enduring mode quickly breaking down and then equally quickly being replaced by a markedly different but equally stable new arrangement is clearly unsatisfactory and historically inaccurate. Such "binary histories" (Sayer, 1989) can perhaps follow too easily from the concept "mode of regulation" (although we do not want to suggest that this is inevitable).

Third, economic history tends to show that the story of capitalism has been one of almost constant upheaval, crisis, and conflict. Periods of relative stability, such as might be labeled modes of regulation are rather rare (and when they do occur, brief) phases (Goodwin & Painter, in press). U.K. fordism is often presented as having lasted from the end of the Second World War until the oil crisis and subsequent recession of the 1970s. Given that austerity measures were in place until the early 1950s and clear economic problems were emerging by the end of the 1960s, however, it is arguable whether it makes sense to identify a full and successful mode of regulation as existing outside the period from the mid-1950s until the mid-1960s. Moreover, some defining features of a supposed mode of regulation may not actually coexist temporally for very long. For example, the provision of public services through the welfare state developed in Britain particularly strongly immediately after the war and then again in the late 1960s and 1970s. On the other hand, productivity growth to fund rising living standards was more evident in the 1950s and 1960s. This suggests that the temporal dynamics of a mode of regulation can be quite complex and can undermine the implication that the term arguably carries of a stable formation that is quickly realized in full and then endures for a relatively long period.

Fourth, we want to suggest that the concept of mode of regulation tends to emphasize structure over strategy and form over practice. The implication in some accounts is that conflict and human agency come into play only during brief moments of crisis *between* modes of regulation. Though strategic action may be particularly significant during periods of heightened crisis and rapid change, it does not disappear between them. Because, as we have pointed out, temporal dynamics exist within any set of regulatory relations, practice should, we believe, be kept center stage throughout the account. This is not to deny that some regulationist texts recognize that structural forms institutionalize conflict within certain strategic rules of the game and accept that research must examine both structure and strategy. In general though, the tendency has been to highlight structure and form.

Fifth, we would argue that the search for a stable mode of regulation has, implicitly at least, led researchers to concentrate their analysis at the nation-state level. This is because the key mechanisms and processes that have been identified as contributing to such stability all operate at the national level. Indeed, the core defining features of fordism as defined by Jessop (1992a, p. 59)—a degree of correlation between wages and productivity and price increases; a role for the state in supporting mass consumption through the social wage; and state intervention to manage aggregate demand—could be success-fully implemented only by the nation-state. Viewed in this way, the notion of a local mode of regulation becomes a contradiction in terms. If we understand regulation as a process, especially as a fluid and uneven process, however, we can analyze the variability of this process. Such analysis of course can, and indeed should, take place at a subnational scale. It no longer becomes necessary to study the national level, and the local, regional, or even the urban level becomes a legitimate arena of analysis. What we are identifying in these instances is not the existence of a stable or unstable mode of regulation but the development and implementation of a set of regulatory processes, which operate at varying scales.

In large part these criticisms of the mode of regulation concept are well known and have already been debated among regulationists and between them and their detractors (Brenner & Glick, 1991; Clarke, 1988, 1990). Although they seem to be cogent criticisms, paradoxically our purpose in rehearsing them here is not to suggest that they fatally undermine the regulationist project. We believe that these critiques of the concept of mode or regulation are just that: critiques of the concept of the mode of regulation. We do not think they fatally undermine the regulationist project because we do not believe that project stands or falls by the notion of the mode of regulation.

### Regulation as Process and Practice

The mode of regulation concept was advanced to answer the central question, posed by regulation theory understood as a methodological approach. That is, how is the reproduction of capitalist social relations secured and developed given its inherent tendencies to crisis and instability? Even if the mode of regulation notion were abandoned, this central question would remain. Our approach to answering it emerges from thinking about one further problem of the concept of mode of regulation. An unfortunate connotation of the concept is the implication that at any one time there must be either "perfect" regulation (during a mode of regulation) or no regulation at all (during an intermediate crisis phase).

Yet this is clearly absurd. Very rarely, if ever, does regulation cease altogether (civil war accompanied by the complete collapse of state institutions is perhaps an example). Equally, even during periods of sustained economic growth, regulation could hardly be described as perfect. The system is simply too complex and the process of regulation too contingent. Any period of stable development will have setbacks and conflicts. Most of the time therefore, regulation is neither perfect nor wholly absent. Rather, it is more or less effective, depending on the mix and interaction of the various factors involved.

We view regulation as a process, rather than as a series of different modes. Instead of looking for coherent modes of regulation, we prefer to emphasize the ebb and flow of regulatory processes through time and across space. At certain times and places, those processes will be more effective than at others. The process of regulation is the product of material and discursive practices that generate and are in turn conditioned by social and political institutions. This challenges the view of the history of regulation as marked by stable and coherent phases separated by brief but sharp discontinuities.

In our account, regulation is tendential, not achieved or established. There are tendencies toward effective regulation that are partly, but not entirely, the products of strategic action undertaken by individuals and institutions. Counter-tendencies operate to disrupt the reproduction of the capitalist order. The regulatory process is the product of the interaction of these forces.

A regulation theory that treats regulation as process is able to deal rather more subtly with temporal and spatial variability. Because regulatory processes are the product of social practices, they must be understood in relation to the concrete contexts of practice. As concrete phenomena with specific histories and geographies, practices must be understood as intrinsically unevenly developed. In other words the *geography* of regulation is not an optional extra or final complicating factor. On the contrary, the process of regulation is *constituted* geographically. Its unevenness is inherent. Indeed, as a product of the contingent interaction of disparate practices, each with its own geography (and history), the process of regulation has a highly complex geography.

This leads us to thinking about the generation of regulation (or, conversely, processes that undermine regulation) as organized in and through key *sites* of regulation. All social practices are situated in space, as well as time, but space is not an abstract container of such practices. Instead, space enters into their production. Each set of social practices that make up the interactions that we identify as regulation has its own key sites, and those interactions are thus interactions across space. The sites themselves will vary in form and process. Some, for instance, will be situated in economic space (e.g., the labor process, mode of growth, international regime), some in political space (local, regional, national, and supranational), and some in social space (e.g., innovation milieu, new industrial district). An urban regime as a site of regulation could be viewed as being situated at the intersection of political, economic, and social space, involving as it does the combination of (local) state capacity with a variety of nongovernmental resources (some social, some economic). An investigation of these varying sites could well offer a fruitful avenue of research within the regulationist perspective. In any case we want to suggest that the spatiality of regulation is integral to its effectiveness or the lack of it. To briefly exemplify what we mean by this, we will consider some practices involved in our own area of interest, urban governance.

Urban governance is produced in and through institutions, including, but not limited to, decentralized state institutions. At the same time, it involves the production of institutions. Practices are institutionalized, whereas institutions are constituted through (partly routinized) social practices. Poulantzas's (1978) account of the institutional materiality of the state captures something of these processes. Because many of the practices of urban governance are situated in institutions, these become sites of regulation. One can see here where the notion of an urban regime may be linked in to these ideas on regulation. The regime itself would comprise many such sites, some governmental, some private, and some deliberately organized around public-private partnerships. The regime itself may even develop new sites and help to institutionalize new forms of social and political practice. Although, by definition, regulatory practices have effects that are spatially widespread, or more widespread than the boundaries of the institution, their effects are rarely ubiquitous: There are sites of resistance and disruption, as well as sites of regulation. At the same time, the production of regulatory processes through social practices can depend on quite particular institutional contexts. Again, both concerns could be investigated through concrete research on urban regimes.

For example, organizational cultures can vary quite markedly between institutions. Central government departments have different institutional understandings of their political world from those of elected local authorities, which are different in turn from those of unelected agencies such as the British Urban Development Corporations. Marked variations also exist between individual institutions within the same group. Even where regimes operate to combine such practices and thereby generate effective regulation, we would resist the temptation to identify a homogeneous mode of regulation. The regulation process is highly differentiated over both time and space. Such differentiation need not necessarily be dysfunctional for regulation at higher spatial scales. It is in principle possible that spatial differentiation can actually help the process of regulation.

As a final note on the mode of regulation concept we should say that our doubts about it stem in large part from the discursive connotations it has acquired. It seems to us eminently possible to rework the notion to consider a view of regulation as process, unevenness, tendency, and practice. Moreover, as Jessop (1990a) has pointed out (see p. 154), the pioneer regulationists had some of these issues, especially those of conflict and human agency, at the heart of their concerns. Rightly or wrongly, though, the notion of mode of regulation has become associated in the literature with stability, coherence, functionality, and structure. These connotations vitiate the usefulness of the term for our research, which is concerned above all with the geographically and historically uneven emergence of regulatory and antiregulatory institutions and practices through social and political conflicts.

# State, Governance, and Uneven Development

## Regulation and Reglementation

States are heavily implicated in the historical process of capitalist development. The presence of contrasting policies and politics in different states, as well as the development of relations between states, is central to the uneven development of capitalist social relations. States pre-existed the emergence of capitalism (at least in Europe) and even distinctively modern states emerged no later than capitalist social relations. In addition states have been subject to their own trajectories of historical formation. Although these have clearly been closely interlinked with the development of capitalist relations, they have not in any straightforward sense been dependent on them. In our view, therefore, there is no reason to accord any explanatory priority to either states or capitalist relations.

During the 20th century, although state policies have not been *determined* by the economy, they have been increasingly directed toward economic affairs. Economic prosperity throughout the capitalist world (and arguably the whole world) has come to represent a key yardstick against which states and governments claim legitimacy. Through policy, strategy, and discourse states have increasingly become central to the fluid process of economic regulation. This is of course one reason for the interest in urban regimes, which have arisen precisely to enable governmental actors to deal with the complexities of managing and securing economic development by cooperating with those from outside. Again, one can see how, in concrete research, an interest in regulation can guide one toward an investigation of urban regimes. Bob Jessop (1995b) has pointed to a key distinction in French between *reglementation* and *regulation*. In French, reglementation refers to regulation in the sense of rule making (in this case by the state), whereas regulation refers to regulation in the regulation theorists' sense of contingently emerging regulatory effects.

In many cases, state policy has involved direct economic rule making. Such rule making can be regulatory in the sense of promoting expanded reproduction of capitalist relations. On the other hand, this rule making can be non- or anti-regulatory. In particular, states frequently address social and moral agendas that could involve policies that run counter (sometimes deliberately so) to the deepening and broadening of capitalism within the national economic space. Conversely, state policy in noneconomic fields can have (often unintended) benefits for the capitalist system. One example of the latter is the founding of the British National Health Service. This was bitterly opposed by fractions of capital at the time, but it socialized many of the costs of reproducing a healthy workforce and thereby lowered the socially necessary labor time and generated a short-run tendency toward increased profitability.

## Uneven Development

Our earlier argument that regulatory processes are geographically constituted and thus inherently unevenly developed applies also to the institutions and social practices of the state. In their government and administration, all states (with the possible exception of small city states) exhibit a degree of internal territorial differentiation. In some cases, this is the result of patchy administrative coverage; in others, it can be caused by uneven levels of service provision. Bureaucracies have their own dynamics of development as do stratified elements within bureaucracies. Thus, any large and differentiated bureaucracy with territorial responsibilities, such as a modern state, has inherent tendencies to geographical variation in its activities. This is notwithstanding the 20th century tendency for governments (at least in liberal democracies) to adopt strategies of territorial evening-out in service provision and administrative competence. Such strategies are often required to counter tendencies to the uneven development of state, as well as private sector, activities. Thus, in Britain, for example, there are regional variations in judicial practice, health service provision, government research and development expenditure, and transport policy, to name but a few.

Most states larger than a certain size have developed with a lower tier of government or administration. In part this may be an attempt to ensure uniform coverage through administrative decentralization. In part too, it allows an explicit recognition of the significance of uneven development through political decentralization: allowing a sensitivity on the part of local agencies of the state to local circumstances and political processes. In some cases, where it is based on ancient structures of local rule, the local government system can continue to bear the signs of its independent local histories. In all cases, however, once they are established, local state institutions will experience their own trajectories of development. Even in a highly centralized state, such as the United Kingdom, considerable geographical variations exist within the local government system.

As Michael Mann (1984) has shown, the modern state has a characteristic tendency toward territorial centralization. That is, although centralized, the modern state attempts to exert its authority across the whole of its territory, in many cases through a system of local administration or more or less locally accountable local government. At present, states are the only forms capable of expressing authority comprehensively across the whole of their territories. At the local level, however, a variety of other institutional actors can also be in a position to make claims to authority. Such claims can be against the state, in cooperation with the state or unconnected to the state. They can be made explicitly or implicitly. Examples of the institutional actors involved include firms, voluntary organizations (NGOs), and arms-length parastate institutions, such as the so-called Quangos in the United Kingdom (quasi-NGOs). In Britain, the Economic and Social Research Council's (ESRC) *Local Governance Research Programme,* of which our research is a part, is based on the assumption that the range of nongovernmental actors involved in the reglementation of local areas is increasing, as are their powers and influence.

### Government and Governance

Within Britain, the implication is that at the local level there has been a decline in government (the role of directly elected local government institutions) and the rise of "governance"—the exercise of authority by nongovernmental institutions coupled with claims (no more than that) to legitimacy.

With the shift from government to governance, studying the uneven development of regulation becomes more complex. Instead of the limited (though differentiated) set of institutions making up the state, we are now faced with a highly diverse range of institutions of many different types, origins, and histories. As certain types of authority are delegated to the private sector, NGOs, and Quangos, the potential for unevenness clearly increases. Indeed one of the British government's legitimating arguments is that this transfer of powers promotes a sensitivity to local circumstances. The network of regulatory institutions and practices has grown more complex and interconnected. Again, we can make the connection with urban regimes. These are one concrete expression of such networks, involving a range of actors from a variety of organizations coming together to promote and enhance institutional capacities in a particular urban setting.

Regulation theory is potentially particularly well placed to investigate these changes. Regulation theory emphasizes regulation over reglementation and has always been concerned in principle with a diverse range of intersecting regulatory forms and processes. There are also some superficial similarities between the concepts of governance and regulation on the one hand and government and reglementation on the other. It would be a mistake, however, to elide some important distinctions. Regulation is made up of processes that are often the unintended consequence of activities carried on for other reasons, whereas governance carries some sense of strategic and goal-directed activity. Where those strategies and activities contribute to the stabilization of economic growth, they are regulatory, but strategic activity could be antiregulatory (whether intentionally so or not). Although reglementation is governmental, not all government involves (economic) rule making. The pairs of terms are thus not completely symmetrical.

## Political Discourse and Local Governance

Returning to a regulation theory that is based on social process, conflict, and strategy turns the spotlight onto discursive as well as material practices. In principle, regulation theory has always been concerned with discourse. Regulationist writers (including ourselves) have often repeated, mantralike, the claim that regulation is a social, political, *and cultural* process. In practice, however, substantive regulationist accounts have rarely considered the contribution of discourse to specific regulatory processes or, in the more usual terminology, modes of regulation. Yet there is no great difficulty in building a sensitivity to discourse into regulation theory. Fordism, for example, clearly depended on some key discursive constructions associated with political consensus, the family wage, the limits of collective bargaining, the mass consumption norm, and so on. Fordism in the United Kingdom was closely connected with car advertising, the ideal of suburban living, one-nation politics, and, later, a discourse of technological advance. The comparative neglect of the cultural and the discursive within regulation theory is unfortunate because regulationist accounts were clearly weaker as a result. Yet such neglect was also unnecessary. Even in its more structural form, regulation theory could have incorporated the role of discourse, perhaps through something like Raymond Williams's (1961) concept of "structure of feeling" (p. 64). Indeed, Gramsci (1971), in his pioneering essay on the introduction of fordism into Italy in the 1920s, described how the new methods of work were inseparable from new ways of thinking and feeling life (p. 302).

We do not have space here to investigate the reasons for the lack of attention to the cultural and discursive moments of regulation hitherto. Suffice it to say that our reading of regulation theory as process grounded in practice cannot afford to perpetuate this neglect. As the possibility of identifying coherent and stable modes of regulation recedes, a study of regulatory processes in local governance must highlight political practice, strategy, and conflict. This leads (or should lead) to an intensified concern with discourse because practice, strategy, and conflict are inherently, though not exclusively, discursive. Indeed, one could investigate and compare urban regimes partly in terms of the discourses that they employ (see, for instance, Stoker & Mossberger, 1994, who identify symbolic or image-building regimes).

This does raise a methodological problem, however. As we have shown, regulation occurs when several relatively autonomous social processes interact to promote relatively sustained and stable capital accumulation. That is, regulatory processes are defined *in terms of their effects*. The methodological difficulty is that the effects of strategies and discourses are notoriously difficult to gauge. Even where a political strategy appears to have achieved its goal, it is often impossible to tell whether the effect was the result of the strategy or a happy coincidence. Further, social systems are subject to often considerable lags between cause and effect. Thus, the evidence for effective regulation can be judged only with hindsight. During a phase defined by conflicting strategies it is certainly impossible to predict regulatory outcomes with any certainly.

A related, but slightly different, difficulty stems from the explanatory structure of regulation theory discussed in the earlier section in this paper on regulation theory and methodology: In reversing the causal arrow (or at least converting it into a double-headed one), regulation theory provides the means for explaining the regulation of the economy (or the lack of it). By extension, therefore, the local state and local governance are among the things that can help explain the regulation of the economy (or the lack of it). The explanation of changes in the local state and local governance *falls outside the scope of regulation theory*. This is perhaps somewhat crude because, as Bob Jessop (1990a) has pointed out, objects and process of regulation are mutually constituting and emerge together (pp. 185-186). We would certainly accept that the local state and local governance cannot be fully understood outside their roles (positive and negative) in the ebb and flow of regulation. Our argument, however, is that neither can they be fully understood within them. The institutions and practices of local government have their own histories and patterns of development. Explaining their changing character thus requires a theory of governance, a theory of the state, and empirical historical and geographical research, as well as a theory of their impact on (economic) regulation.

What is required, therefore, is an investigative frame that joins a geographically sensitive regulation theory based on social processes to a critical political sociology of the local state and urban governance based on an investigation of the material and discursive practices in which the local state and urban governance are grounded. The crucial question that follows from this is whether a reconstituted regime theory can contribute to this critical political sociology of the local state. We have tried to indicate several spaces within a regulationist account where regime theory might fruitfully be employed. Addressing this question more fully lies at the heart of the next part of the book.

## Conclusion

We have argued that at the present time, work within the regulationist framework needs to prioritize concrete research. We have two main reasons for emphasizing the role of concrete research within a regulationist approach. First, we regard context as central to understanding social practice and practice as central to understanding regulation or lack of it. Second, grounding an analysis of regulation in the concrete highlights the often complex geographies of regulation—which are themselves crucial in understanding the development of the practices of regulation.

Regime theory helps to pinpoint particular aspects of such regulation and highlights their operation within an increasingly complex system of urban governance. Regime theory is especially useful in pointing to those new forms of networks and partnerships that have emerged as urban government has gradually transformed into governance. Regime theory is also helpful in drawing attention to the increasing intervention of local governments in an economic sphere that was previously seen as the province of the private sector. This kind of research on urban regimes focuses attention on institutional and practical *responses* to what we understand as the failure of fordist regulation and on the role of uneven development therein. We believe it is important to view the actions of various urban regimes in this light and to appreciate how the concerns of regulation theory can help us conceptualize the organization and actions of urban regimes.

# Notes

1. We are using the English translation of a book originally published in French (Boyer, 1986).

2. In our view, although considerable work has been done on the temporal dynamics of capitalism, consideration of spatial variation, though still significant, has been less systematic. One concern of our work on the regulation approach is to enrich further regulationist accounts of space and geography.

3. This applies, of course, where the object of regulation is defined as the economy. If one were to consider the state, for example, as an object of regulation, then the converse would presumably apply.

C
H
A
P
T
E
R

# 3

# Spatial Structures of Regulation and Urban Regimes

MARSHALL M. A. FELDMAN

W e cannot create a more perspicacious urban political economy simply by combining urban regime and regulation theories. They differ in their levels of abstraction, spatial scopes, purposes, and objects of analysis. Regime theory is about "development politics" (Stone & Sanders, 1987) and how governing coalitions, which businesspeople often dominate, shape local urban development policy. Regime theory might therefore be better labeled economic politics than political economics. Regulation theory, in contrast, focuses on broad epochs in capitalist history and the large-scale regulatory processes,

Thanks to Rich Florida, Mark Goodwin, Andy Jonas, Mickey Lauria, Joe Painter, Rolf Pendall, Erik Swyngedouw, and Dick Walker for their comments on this and related work. Special thanks to Ric McIntyre, who, as colleague, critic, and friend, helped develop my thinking on this subject. Alison Nisbet and Nuaqi Yuan worked as graduate assistants on the original research and helped shape the questions here. I gratefully acknowledge U.S. Economic Development Administration Award #99-07-13734, National Science Foundation Grant #SES-9122778, and Rhode Island Agricultural Experiment Station Grant H-707 for supporting this research. The usual disclaimers apply.

including a society's "culture, manners, myths, and dreams" (Barbrook, 1990, p. 92), that account for "the variability of economic and social dynamics in space and time" (Boyer, 1990, p. 27).

Yet both have an overarching concern with *governance.* Regime theory emphasizes how dominant political coalitions establish and maintain the capacity to govern (Stone, 1993), whereas regulation theory emphasizes the governance of production systems (e.g., Storper, 1991; Storper & Harrison, 1990). So integrating these theories has considerable potential, but several thorny problems stand in the way. Regime theory inadequately theorizes connections between local agents and their wider institutional context, whereas regulation theory underestimates the importance of local actors and institutions. Neither does a particularly good job with different spatial scales. Regime theory usually ignores metropolitan or larger spatial scales, and regulation's theory broad conceptual apparatus gives short shrift to spatial variations in material and discursive practices and their relation to accumulation regimes' relatively unified logics.[1]

Perhaps most damaging, both theories depict capitalism inadequately. Regime theory has virtually no explicit theory of capitalism except to claim the liberal state often must deal with business to enact policy. Consequently, regime theory frequently degenerates into ad hoc empiricism (Cox, 1991b; Cox & Mair, 1991).[2] Regulation theory, on the other hand, often reduces capitalism's complexity to discrete transformations between homogeneous accumulation regimes, thereby ignoring spatial and industrial variability and more gradual yet irregular diffusions and transformations of social practices. It tends to overlook how material practices of everyday life constitute modes of regulation and acts as if modes of regulation are legislated rather than constituted through practices situated in dynamic local settings of experiment and conflict. Regulation theory also lends itself to overemphasizing temporary equilibria between supply and demand while ignoring other, fundamental aspects of capitalist accumulation and crises (Walker, 1995).

This chapter aims to fill in some of these missing links (Tickell & Peck, 1992) between the local and global in regulation theory and between regulation theory's political economy and regime theory's economic politics. I propose a theory of *spatial structures of regulation* as a fundamental retheorization of geographic industrialization under capitalism. Capitalist production necessarily involves qualitatively distinct flows and relations, which I refer to here as processes, all of which are necessary conditions for ongoing capitalist production. These cannot be reduced to input-output systems or to simple systems of governance or regulation. Each has its own ontological status with a distinct spatial scale, pattern, and dynamics, or, more generally, its own "spatiality"

(Soja, 1989). A spatial structure of regulation (SSR) is a particular combination of these processes. A local economy's dynamics derive from its SSR and reflect the logic stemming from the combined interaction of the constituent elements. We must understand the space economy as the complex workings of these elements as various means of regulation articulate them to and coordinate them with each other, as they express their internal contradictions, and as they contradict each other and are struggled over in concrete production systems.

SSRs, and their spatial organizations, are social creations. Each process embedded in an SSR has its own spatial scale, and the processes mutually condition each other. They may impose a certain scale among the others, substitute for others at a particular scale, or preclude or hinder others from having certain scales. Moreover, spatial organization and scale usually are emergent properties of SSRs rather than intended or optimal outcomes. Hence, spatial scope and organization are highly contingent, concrete properties of any given SSR. Consequently, the local state plays a distinct and contingent role in the SSR.

This chapter has three parts. The first identifies several processes in contemporary capitalism. The basic issues here are the ontological status of each separate from the others, factors governing their spatial forms, and their key distinguishing features. For the sake of brevity I downplay perhaps their most important features—their internal dynamics and contradictions, their reproduction, and their causal efficacy. A discussion of means of regulation follows. Means of regulation are abstract conceptions of the ways capitalism's processes *can be* regulated and constituted. The third and final section sketches out how one can apply this framework to integrate regulation and regime theory. By identifying means of regulation we can identify the different ways regulation *might* take place. We can then examine their prevalence in local economies and ask how they come into being. This also can lead to identifying *alternatives* to mechanisms embedded in the state as a regulatory institution and shed light on roles varied urban regimes and their policies play.

# Production Processes

Any stable economic system with a developed division of labor involves multiple processes.[3] These all involve flows between social entities and therefore involve *social relations.* Moreover, because they exist in and as space, they are inherently spatial. Not all such processes can be found in all modes of production—some are unique to capitalism and others have a particular capitalist form.

Regulation theory centers around such processes and their distinctiveness under capitalism. We can identify at least six processes essential to capitalist production: materials, value, personnel, information, property rights, and authority. Each has its own social, spatial, and temporal pattern, which may coincide in full or in part with one or more of the others'. For example, value and materials processes overlap where money is exchanged for a physical good. But because money can also flow where no physical goods are involved, as in interest payments, the two processes coincide only partly. In addition, some processes presuppose others or are mutual preconditions for each other but never coincide. For example, legal property rights supported by the state define the entities between which value flows. Here I examine such flows, the social relations they entail, and their spatial characteristics.

## Materials

Discussions of capitalist economies usually treat production as the physical transformation of nature because physical materials are central to most capitalist production. Whereas most neoclassical analysis could take place on the head of a pin, the spatial economics descendent from Alfred Weber and von Thünen emphasizes the physical flow of goods through space. Recent Marxian work also reflects this concern. Sheppard and Barnes (1990) note that transportation is part of production so that an area's transportation system affects local output and the structure of prices and values. Swyngedouw (1991) takes this one step further, arguing that the physical flow of goods depends on spatial organization so that spatial organization is itself a force of production.[4]

Flows of physical material are central to the nexus of concerns bound up with input-output systems, flexible manufacturing, and economies of scale and scope (Scott, 1986; Storper & Harrison, 1990). For example, the flexibility and efficiency of material flows are believed to give competitive advantage to both the Japanese *kanban* system of supplier linkages and the new industrial spaces of vertically disintegrated flexible specialists. Understanding material flows in production processes is essential for understanding these new forms of spatial and competitive organization (Walker, 1989).

Production processes differ for different materials: assembling jet engines is very different from finishing fabrics. Different materials not only involve different processes, but the technologies pertaining to different materials advance at drastically different rates (Walker, 1989). Scientific advances typically revolutionize processes, pertaining to one type of material process while leaving others relatively untouched. Chemical and electrical technologies advanced dramatically in the late 19th century, mechanical assembly led the way in the early 20th century, advances in electronics and information technologies are revolutionizing the current period, and biological technologies will likely play a key role in the 21st century. Because such radically different materials-related processes have different implications for production topologies, they also have different implications for production's spatial, social, and institutional organization. Therefore, labeling accumulation regimes in terms of a form of production (e.g., fordist assembly lines) is misleading because sectors dealing with different materials always differ.

Materials flows involves the movement of goods, many of which are bulky, over long distances. These flows are therefore intimately bound up with physical infrastructure: canals, roads, railroads, airports, and so forth. Historically, physical infrastructure has been associated with large capital investments and therefore closely tied to valorization and the state (Feldman, 1977). In addition, infrastructure's physical fixity can itself be a fetter on flexibility and a constraint on future possibilities. When it found itself in a new competitive environment in the 1970s, the United States was at a competitive disadvantage because of its relatively costly and energy-intensive spatial form (Feldman & Florida, 1990). If flexibility is indeed becoming more important in competition, then sectors and places using materials with heavy requirements for fixed infrastructure are in a far more precarious situation than those specializing in processes less tied to infrastructure. For this reason if no other, services and similar sectors can respond more rapidly and flexibly to economic instability than do "heavy" industries such as steel.

Material processes also have direct "environmental" consequences, as a drain on resources, as a disposal and recycling problem, and as a threat to health and ecological systems. Practices aimed at ameliorating these consequences shape the SSR. For instance, recycling may lower aggregate demand. This relation can be subtle and spatially varied. Nelson (1995), for example, found a correlation between states with weak growth management and per capita rates of bank failure. Evidently, in this instance, local attempts to ameliorate environmental impacts caused by material growth also stabilized the local financial process by protecting financial institutions from their own propensities to speculate.

## Value

Regulation theory often builds on Marx's reproduction schemes (cf. Lipietz, 1986, to see the implicit, but tight, relationship between regulation theory and Marx's reproduction schemes). Marx's theory is actually a "dual" theory of commodities, prices, and values (Sheppard & Barnes, 1990, p. 43). If a commodity's labor value is the socially necessary labor time to produce it, labor values are shadow prices (Morishima, 1973). Marx demonstrates the *possibility* of stable accumulation with multiple sectors, and regulation theory addresses the historical conditions that realize this potential. For the purpose of exposition, Marx assumes investment flows from branches with low profit rates to those with high rates until adjustments in supply equalize profit rates, but this need not be the case. For example, different profit rates can prevail in different branches of a segmented economy. Actually, the notion of equilibrium prices runs counter to how real capitalist economies work because capitalism's dynamism continually revolutionizes the structure of value (Walker, 1988).

Capital's "circuits" are more complex when we consider space.[5] A commodity's value depends on *where* it is produced (Sheppard & Barnes, 1990). If value is best-practice labor time, valorization depends on demand because the best-practice technique for one level of output may be entirely different for another. Variations in conditions of production and demand imply that what is best practice in one locale may be different in another. Moreover, capitalists can "satisfice" by accepting less than maximum profits, a practice that depends partly on spatially varying custom. Despite globalization, many commodities still have a local character, and valorization occurs within a limited spatial range depending on production technology and local custom. As spatial differences are rearranged through "space-time compression" (Harvey, 1989), previous configurations of circuits of commodities, prices, and values are revolutionized. Economic restructuring therefore involves not only corporate strategies and sectoral and spatial shifts but also drastic realignments of relative prices and investment in speculative arenas and seemingly safe harbors as capital seeks shelter from restructuring's storm (Feldman & Florida, 1990).

As social relations, the value circuit has traditionally been the terrain where classes are defined. Here two things are evident (Sayer & Walker, 1992). First, insofar as the circuit has distinct spatial form, class relations must vary in space. Class conflict, exploitation, and so on are as spatially variable as the circuit that defines classes. Second, because the value circuit is not the only important circuit in advanced capitalism, it is not the only structural axis for conflict. Capitalist societies have other dimensions of social conflict, such as race, gender, ethnicity, and nationality, but these have proven difficult to integrate into a more general theory of capitalism. The theory presented here may or may not prove useful for this, but it does raise the issue of social conflict along several other dimensions integral to capitalism.

The circuit of finance capital is related to, but not identical with, that of value (Harvey, 1982). Money, as a store of value, can bridge uncertainties in space and time (Clark, 1981b). Economic agents know they are uncertain about the future economic environment and their competitors' and customers' actions. Agents therefore act strategically, adding to overall uncertainty, and disequilibrium conditions prevail. Because it acts as a buffer, money itself is in demand as more than mere representation of value.

Finance capital has garnered considerable attention recently as it expanded without a concomitant expansion of value (i.e., as debt grew; see Harvey, 1989), international finance became an important industry in its own right (Thrift, 1989), and access to finance capital became increasingly important to regional growth (Florida & Kenney, 1988). The latter work on venture capital is especially relevant here because it demonstrates the distinctly spatial character of the financial circuit. Venture capital originates in some places and flows elsewhere. Furthermore, equalizing profit rates do not govern such flows; quite the opposite, distinctly regional practices of competitive behavior govern *monetary flows,* and such flows *disequilibrate* profit rates between regions.

## Personnel

Stable accumulation requires periodic activity systems that render both the availability of labor and consumption patterns predictable. Although migrant workers are still common and markets for certain professions are national or even international, most labor markets in contemporary capitalism are organized around daily activity systems. Generally, such systems involve journeys falling into one of four categories: to work, to market, at work, and other journeys of reproduction. Ever since housing markets developed into a sphere distinct from employment, the journey to work has been an important organizing element in capitalist urbanization (Feldman, 1977). The journey to market for purchase of items for final consumption has grown in significance, first with the advent of fordism and mass consumption and, more recently, with growing use of consumer services and varied, "postmodern" consumption patterns. The journey *at* work is also apparently growing in significance and calls for much more attention. Interchange of personnel between firms is said to characterize both flexibly specialized industrial districts and the Japanese kanban system. Business travel is becoming increasingly important, and access to an international airport with direct flights is sometimes as important in industrial location as availability of suitable labor in the local labor market (Malecki & Bradbury, 1992). Finally, there are trips (to schools, parks, etc.) that should be thought of as essential components of an SSR. For example, the location of child care plays an important role in labor supply and family demographics (Van Allsburg, 1986).

Migration, when not part of migrant workers' periodic movement, is a change from one location to another and from one periodic activity system to another. But migration is not random, and some areas are more involved in migration than others. An area's place in migration patterns directly affects its economic fortunes. This was as true for 19th century New England, whose textile industry relied on technologies and labor imported from Britain, as it is today for the sweatshops in Los Angeles and New York, which rely on migrants from Asia and Latin America. Moreover, labor's embeddedness (nonmigration) can impede capital mobility through worker resistance to industrial restructuring and, in general, makes labor a local actor (Cox & Mair, 1991). Thinking of migration as strictly the *effect* of economic development and purely a response to relative prices is a mistake. Migration must be socially constructed, and all areas do not have the same access to migrants.

The personnel circuit exists in another sense, in that the personnel themselves are socially "created" through social reproduction. Somehow, through social life, labor's personality is re-created as attitudes, meanings, institutions, and practices (Jonas, in press-b). A good deal of recent attention focuses on "local culture" as a source of spatial difference (Duncan & Savage, 1991). Although this opens up new vistas regarding the local construction of meanings and economic development, it is important to note that culture extends beyond symbols, meanings, and attitudes to encompass practices and institutions. These give an objective structure to social life and encourage some sets of meanings and attitudes while discouraging others. A very complex web exists here that needs unraveling.[6]

## Information

Recently, there has been more appreciation of information's role in capitalist development (Castells, 1991; Hepworth, 1990). Information is too complex to be reduced to price signals, nor can it easily be a commodity. Its usefulness need not be reduced by its consumption and can even be enhanced—witness computer manufacturers' proclivity for setting "open" standards and putting certain software in the public domain. Furthermore, information's usefulness often cannot be known without the information itself (Arrow, 1971), and information production is a social process in which shared knowledge is essential to the production process itself: the free flow of information contributes to the quality of information produced.

Still, information does not "fill the air" willy nilly. It is transmitted and has both a source and destination. One can distinguish between two kinds of information. *Routine information* is repetitive and pertains to events already embedded in the structure of everyday life, for example, the price of pork bellies, the location of an oil tanker, or the balance on a credit card bill. Nonroutine, or *innovative,* information pertains to events not yet regularized in everyday life.

Innovative information becomes particularly important during economic restructuring, when new innovations may have increased likelihood of implementation (Mensch, 1975; Schumpeter, 1939).

New information technologies allow quick transmission of information over long distances, but only routine information can take full advantage of this. Routine information typically is *gathered* and *processed,* whereas innovative information is *produced.* Routine information can be processed in a wide variety of locations, such as in a suburban back office or offshore, and management can control routine information workers rather tightly. Innovative information requires openness: workers producing innovative information *must* have freedom, and they are more productive when located where they have a high likelihood of receiving new information. Although such information may be transmitted through computer networks and the like, face-to-face contact is still most effective. This partly explains the agglomerative character of some innovative industries and the "seedbed" function traditionally attributed to urban areas.

Unlike product cycle theories that blindly attribute a seedbed function to dense urban areas, thereby reifying population density, a theory of SSRs must consider the actual functioning of such systems. The social regulation of information flows differs in different areas. In some locations corporate structures hold information tightly, whereas in others information is much more free flowing in what is sometimes mislabeled an "entrepreneurial climate." In some SSRs, an elite group of engineers and technicians monopolize innovative information, whereas in others information is widespread and flows up, down, and across organizational hierarchies (Saxenian, 1994). A locality's social regulation of information is key to understanding its potential for competition through innovation.[7]

## Property Rights

Buying and selling commodities involves transferring property rights, but property rights can be transferred without commodity exchange. Moreover, one must distinguish between rights of ownership, possession, and control (Carchedi, 1977). Broadly defined, property rights delineate the modern corporation's boundaries. Therefore, distinguishing flows of property rights from capitalism's other flows is absolutely essential in grappling with vertical integration and disintegration. For instance, disintegration with regard to property rights can simply spread risk so that flexibility in materials flows within an agglomeration of vertically disintegrated firms could just as easily be achieved under the aegis of a single corporation. Interpreting such an agglomeration as efficient from the standpoint of production costs would therefore be a drastic misreading.

Property rights delimit obligations and responsibilities, and their transfer may be a way to circumvent such constraints. Because nation-states define property rights, multinational strategic alliances that blur boundaries between national firms can circumvent protective national restrictions. Blurred property rights also blur the boundaries of traditional input-output systems. For example, the textile industry is commonly organized through "convertors"—firms that organize production, from weaving the original cloth to delivering the final product. These firms can purchase the raw materials and own the fabric throughout the production process, which can span thousands of miles and involve several distinct enterprises. In an input-output sense, this is vertical integration: the converter owns the fabric from start to finish. From the standpoint of material flows, however, it is vertically disintegrated: every step involves a separate firm. Nisbet (1991) reports on one textile finisher that is a wholly owned subsidiary of a convertor but does about 70% of its business for other convertors. So about 30% of its business is organizationally integrated, and the remaining 70% is organizationally disintegrated: only ownership (property rights), rather than any other characteristic, distinguishes the products produced under integrated and disintegrated production. The entire situation is indecipherable without explicit consideration of property rights.

## Authority

Authority pertains to power over others and to the power to take certain actions (e.g., to invest capital). Authority can come from several sources and can be delegated. In this sense authority is a flow. Authority can attach to ownership of capital or to a particular place in the social division of labor, so it can flow without conscious delegation. Moreover, the delegation of authority does not necessarily alienate it from its original owner: the CEO who delegates authority can reserve the right to override decisions.

Authority is frequently fluid and open—terms of authority can be negotiated and interpreted, and when asserted over others, it often depends on their compliance. Authority is sometimes granted by the party over whom authority is exercised: things are very different when the worker accepts the boss's authority than when the worker tries to undermine it. Parties to a contract can look for loopholes and other ways to undermine each other's authority. Spatial location can be used to contest or to undermine authority (Clark, 1981a). Because spatial organization itself varies, spatial location's effectiveness for this purpose itself varies in space. Furthermore, in a decentralized legal system, such as the U.S.

federal system, the legal boundaries of authority relations can vary in space. For these reasons, authority relations have distinct spatial patterns.

## Means of Regulation

Processes pertaining to materials, value, personnel, information, property rights, and authority are essential to contemporary capitalism, but their importance varies. If the materials or value processes fail, a crisis results. On the other hand, if all capitalists suddenly stopped innovating, logically the system might still function. Some processes can be interrupted for a period without precipitating a crisis. For example, the state can disrupt the finance process temporarily in a bank emergency without halting the entire system. Also, the processes are distinctly *spatial,* and their functions are duplicated in many places. So a process's failure in one place need not precipitate a general crisis. A locality whose transportation system is so overloaded that workers cannot get to work may be in big trouble, but the overall system may scarcely feel the effect. The spatial division of labor is itself a source of stability for the system as a whole.

Markets cannot coordinate all six processes with each other or internally (Jonas, 1996; Walker, 1989), so the processes must be *regulated* for stable accumulation to occur. It seems essential to distinguish between the processes themselves and the means for regulating them. Moreover, the processes are always spatial, so it will not do to distinguish spatial means of organization from others. Also, we should distinguish means of regulating the processes from the specific social and institutional forms that use those means. Furthermore, traditional economic categories may be inadequate for considering such things as the social regulation of information flows. We therefore need some general conceptions of how social relations can be structured to regulate the various processes. At a minimum, five "means of regulation" can be identified: command, exchange, reciprocity, altruism/solidarity, and custom.[8]

In applying these categories, we must avoid functionalism. Transaction cost theories of vertical integration and agglomeration, for example, explain the choice between firm (command) and market (exchange) by their respective functional advantages (e.g., Piore & Sabel, 1984; Scott, 1986). Human societies are not rational creations. We do not find exchange at the supermarket and reciprocity in the nuclear family because markets are better suited to the former sphere than the latter. To be sure, means of regulation are not used in crucial spheres of life if they cannot possibly regulate them. But the fact that a means of regulation is not "optimal" for a specific sphere does not mean it will not be used. To understand why different means are used in different SSRs, one needs to understand how a SSR developed historically and how the different means do or do not fulfill the system's functional imperatives.

## Command

Command involves the ability of one actor to direct another to take specific actions. In general, command calls for *surveillance* to ensure commands are carried out.[9] The spatial extent of command depends heavily on the social and technical means for issuance, enforcement, and surveillance. The factory, a spatially enclosed workplace, developed largely to facilitate management control through command (Burawoy, 1985; Marglin, 1974). As the origin of command becomes removed from day-to-day operations it can grow out of touch with what is actually happening. This is essentially a problem of information: inadequate information flows up from the bottom, and a highly centralized command center has difficulty processing all the information that comes to it. Moreover, delegating command increases the need for enforcement and surveillance because all actors in the chain of command need to be monitored. At the end of the 19th century, hierarchical control proved costly and inefficient for many corporations for these reasons. This crisis of control played an important part in the crisis of the existing accumulation regime around the turn of the century (Edwards, 1979, chap. 4). Indeed, fordism's assembly line was particularly useful for regulating the pace of work without extensive directives and surveillance.

## Exchange

Exchange is common in the circulation of values, prices, materials, commodities, finance, and property rights. It is less common, but not impossible, in the realm of personnel, authority, or information ("I'll tell you my secret if you tell me yours"). Exchange is also bound up with changes between different means of regulation and between processes. Labor contracts, for example, substitute command for exchange. Items from different processes can also be exchanged (e.g., money and information).

Because exchange involves only the immediate concerns of individual parties, there is no need for central planning or any more global intelligence than the parties bring to the table. But command and other means of regulation can also be decentralized and may only involve the transacting parties. Exchange is one of several means for coordinating diverse activities. Furthermore, not all exchanges, even monetary ones, involve markets. Political corruption, for example, requires secrecy and is therefore antithetical to markets, but a payoff for a political favor is still an exchange. Market systems vary, including, for example, private and planned markets (Lindblom, 1977).

Exchange generally has a distinct spatial dynamic. It is somewhat antithetical to communal bonds, and the latter tend to develop within distinct areas. In contemporary capitalism other means of regulation prevail in social units with strong interpersonal bonds, such as the family. When a neighbor borrows a lawn mower, most people do not demand a quid pro quo, but the situation is very different if a stranger from outside the neighborhood asked for it. Neighboring towns frequently have reciprocal agreements for emergency services and the like: individual transactions between towns do not involve exchange, but spatial propinquity makes it likely that all parties will at one time or another receive the benefits of the agreement. Much of the work on industrial districts (e.g., Scott, 1986) claims that some localities lower transaction costs through trust and other relations, thereby embedding exchange relations (rather than command relations, for example) in a place. Conversely, transactions over long distances often involve exchange, and exchange plays a key role in space-time compression (Harvey, 1989). Still, markets require stable institutional infrastructures and often involve face-to-face contact. They therefore tend to be spatially delimited.

The disadvantages of exchange are well known, although the terminology is thick with ideology. Markets "fail" when public goods, externalities, or monopoly power are present. Market exchange also involves considerable transaction costs.[10] Markets require that parties have adequate information—a requirement that is unsatisfied for many complex goods. Markets can be influenced by false advertising and other means of manipulating information. Information itself is difficult to exchange in markets because knowledge of the commodity and the commodity itself are inseparable. Left to their own devices in markets, individuals often make socially undesired decisions. So other means of regulation frequently augment markets.

### Reciprocity

Whereas exchange involves transactions between two parties, reciprocity involves more collective interaction. A computer bulletin board, for example, depends on reciprocity: people contribute information to the bulletin board and take information without a quid pro quo. Similarly, donations to a blood bank are reciprocal relations.

Cooperation can be an important form of reciprocity. Another neoclassical fairy tale is that people naturally separate work from leisure. In fact, people often find work fulfilling and choose to work (Polanyi, 1944). Cooperation combining social interaction and work has its own intrinsic rewards, so even without a division of labor, cooperation can be more productive than the work of isolated individuals. When the extra product is distributed among the participants, this non-zero-sum aspect is further incentive to cooperate.

Historically, reciprocity is more prevalent than exchange. Societies have their norms and obligations that people fulfill to participate in the society or for social approbation, rather than for individual gain or for fear of punishment (Polanyi, 1944). Reciprocity is typically established through norms and based in a territorial community. It is likely to be found in industrial districts, but most researchers have ignored it. How, for example, do norms of reciprocity differ between regions? Do such differences account for differences in development trajectories? How does local policy encourage or discourage reciprocity? These are just a few examples of important questions that research has not addressed.

### Altruism/Solidarity/Consensus/Democracy

Reciprocity still involves a payback. Individuals contribute to the group and get something in return. In contrast, altruism involves no payback. The related concepts of group solidarity and ethics imply a similar giving with no expectation of something in return. People give to charity, join the army, or run into a burning house to save a stranger's child because they believe it is the right thing to do. All human interaction involves belief, so the mere presence of beliefs is not the issue. Instead, the issue is a specific form of belief, how such beliefs are socially constructed, and the role they play in any given SSR.

Altruism and solidarity played an important role in fordism. The welfare state obtained much of its legitimacy by appeals to altruism. Reaganism and Thatcherism struggled to discredit and deflect this altruism. The state was cast as incompetent, and a mythical "thousand points of light" were invented as an alternative means for satisfying altruistic aspirations. At the same time, high fordism's altruism ("ask not what your country can do for you") was attacked as an impractical ideal.

Civic "spirit" does seem to be important in local economic development. This is because altruism and solidarity defined in terms of *place* often lead to concrete actions. Baltimore paid almost $300 million to attract a professional football franchise, and cities across the United States competed to host the World Cup games in 1994. On a much smaller scale, small businesses frequently have strong overtones of paternalistic altruism. Nisbet (1991), for example, describes a 20-person firm whose owner justifies his opposition to unions by citing his flexibility in doing such things as granting individual employees time off when a family member dies. Although he undoubtedly has other motivations, it would be arrogant, disingenuous, and scientifically wrong to dismiss such expressed altruism as pure rationalizations.

### Custom

Custom has generally been ignored in economic development. The localities literature (Jackson, 1991; Longhurst, 1991) and Storper's (1991) work on conventions promise to change that. Any complex form of human interaction involves a host of unstated behaviors and mutual expectations all of which cannot possibly be negotiated or mandated. Custom is therefore implicated in all forms of production. Custom involves aspects of social regulation beyond conscious, intentional actions. Actors follow custom "because that's the way things are done around here."

Custom is long-lasting and local. It is literally built into everyday life. Even where actors know of other ways to do things, custom is often hard to change. Custom involves concrete practices which, to the extent that actors rely on each other, they cannot unilaterally change. If all actors involved want to change custom, doing so can require collective action that is impossible to achieve. Unless there is a social institution to coordinate the change, the individual actors may face a "prisoner's dilemma": the change will not be effective unless all actors make the change more or less simultaneously. Attempts to convert to the metric system in the United States floundered on precisely this dilemma. In most instances, neither the collective consciousness nor the social institutions exist to change custom. Societies therefore "muddle through" from one set of customs to another. In some instances, a well-organized group deliberately changes custom. For example, the "American Dream" of a detached, privately owned house was consciously created by government and business elites over more than a half century (Florida & Feldman, 1988).

# Spatial Structures of
# Regulation and Urban Regime Theory

This outline of a theory of spatial structures of regulation is very partial and incomplete, yet it indicates directions for a radical reworking of regulation and regime theories. Here I will illustrate this by focusing on information and drawing on my own research in New England. I will then conclude with a more explicit commentary on urban regime theory.

*Information* is essential to capitalist production but intrinsically difficult to commodify.[11] Patents are one way to get around this problem: they use the state's *authority* to enforce *property rights* over information. Yet in many situations capitalists opt for other alternatives. Firms subject to intense innovative competition, for example, avoid patents because applying for one requires disclosing information and, in specific institutional contexts, the process takes time. Competitors can use a firm's patent application to discern its strategic directions, to imitate the innovation before the patent is awarded, and to develop close copies that do not violate the patent itself. Moreover, technological rents are greatest during an innovation's early life and quickly dissipate in highly innovative industries. Consequently, such firms rely on internal secrets and proprietary information. Here, management uses *its* authority to declare what is or is not secret, and its *proprietary rights* to such secrets hinge on the *labor contract* with its employees. To protect such secrets, the firm may adopt a strategy of vertical integration, bringing sensitive parts of the labor process under its *authority* and *legal ownership*. Alternatively, the firm can enter into strategic alliances based on a combination of formal contracts (*exchange*) and mutual trust (*solidarity* and *reciprocity*) (see Sayer & Walker, 1992, chap. 3). These are different strategies, in part reflecting different conditions and whose success partly depends on the future course of those conditions, rather than absolute alternatives, one of which is always superior to the other.

These are not the only alternatives: information can be *spatially embedded*. For instance, New England's technological expertise in metalworking can be traced to its early textile and gun-making industries; in electronics, to instruments and electrical manufacturers; and in medical research, to its universities, hospitals, and early public health initiatives (Feldman & McIntyre, 1994; Storper & Walker, 1989). This industrial history underpins the region's dominance in medical instruments, which relies on all three technologies. Yet when New England medical device manufacturers need handmade, precision-crafted surgical cutting tools, they get them from Germany, and when they need low-cost scissors and similar implements for medical "kits," they get them from Pakistan (Feldman & McIntyre, 1994). Each region has its own embedded culture of expertise (precision machine cutting in New England, skilled artisan metal handcraft in Germany, and skilled handcraft in Pakistan) and supporting institutions (community colleges, trade unions, and family networks in New England; an apprentice system in Germany; and family homeworking in Pakistan) that involve the state, at various levels, to differing degrees.

We see by this example how information can be spatially embedded and tied to spatially diverse institutional structures and how these local regulatory structures can combine within a single global industry. Much recent regulationist research makes the unstated assumption that this is an impossible or at best transitional situation. In contrast, a theory of SSRs can help us see the historical unity in such diversity. No one "best practice" eventually outcompetes and pushes out all others. Spatially varied social structures of regulation can be articulated together in a world system whose coherence depends on the uneven development of diverse regional practices. Instead of "flexible specialization" or any other monolithic model of industrial organization, a model in which global institutions integrate radically different local systems into a global system of "nested" SSRs may best capture postfordism's distinctiveness.[12]

Within this framework, the local state plays diverse roles: what can be beneficial to capital in one place may be detrimental in another. The SSR concept might then be fruitfully joined with urban regime theory to provide insights in two senses. First, analysis of a locality's SSR could help identify what issues are important to local capitals and explain their involvement or noninvolvement in local politics. Second, it could help us better understand the impact of local public policy. Regime theorists are fond of claiming "politics matters," but they hardly ever tell us how except in the tautological sense that politics determines political outcomes. A deeper, systematic understanding would allow us to understand what differences, if any, public policies make.

We must radically transform urban regime theory, and in particular its methodological underpinnings, to use it either way. Take deciphering the material structures underpinning political involvement. This resembles regime theorists "latent interests" that

> involves viewing groups as being embedded in larger sociopolitical structures
> and in spatial values; it also involves attributing interests to groups in terms
> of their social roles in these relationships. This makes it possible to specify
> the policy orientation that such groups in all cities would be expected to adopt,
> to predict their actions in terms of these roles and the resources that are
> available to them, and to analyze the conditions under which these interests
> do or do not act as anticipated. (Clarke, 1987, p. 112)

This tacitly invokes methodological individualism. On one hand, it assumes individuals have interests, could know them, and would act on them in the absence of mitigating conditions. On the other, it assumes such individual action can adequately explain political outcomes. In part this becomes a self-fulfilling prophecy because regime theorists define political outcomes in limited ways that relate back to attributed individual interests. Thus, for example, Clarke (1987, pp. 117, 119) points to Dayton, Ohio's emphasis on policies for job retention and creation as evidence that local officials "do not favor more concessionary [to business] business policies," as if "job generation" (i.e., reproduction of wage labor) subverts business! Policies emphasizing job generation in no way imply local policy makers are free from capitalism's constraints or that capital*ism* (as opposed to *individual* capital*ists*) does not dominate policy.

Regulation theory, and more generally critical social theory, aims to understand how structures-in-dominance persist through the dominated's "doxic submission" (Bourdieu, 1994; Lipietz, 1988). Regulation theory's insights come from its ability to decipher accumulation regimes from their "emergent properties" (Sayer, 1992), which are not reducible to individual action or interests. This does not imply, as is sometimes mistakenly supposed (e.g., Stone, 1991), economic foundationalism or determinism, nor does it deny the importance of "agency" in favor of "structure." Regulation theory pays special attention to questions of culture and politics, but it does not *reduce them* to self-interested individual action. Instead it rejects a singular focus on self-interested individual action as leading to *inadequate* social theory.

Second, regime theory often treats groups as unitary bodies rather than as amalgamations of diverse individuals sharing common situations but differing in many ways (e.g., by age, personal history, etc.). Its concept of "interests" is such that if one could put aside this messiness and distill "pure" capitalists or workers in the lab, one could readily discover their interests. In contrast, a theory of SSRs implies that *interests,* if this term has any meaning at all, are overdetermined. They are socially defined, contradictory, strategic, spatial, and not objectively knowable. Whether an action serves one's "interests" cannot be known in advance because the action's outcome depends on the subsequent actions of others. Moreover, people do not act just to further their interests: custom, altruism, command and fear of sanction (actual or perceived), and so on, all play a part. These elements enter into a socially constructed conceptual apparatus through which people filter their experiences. If any interests motivate actual conduct, they are *perceived interests* rather than *objective* interests tied to a sociopolitical role. One main goal of a theory of SSRs is to understand how such perceived interests are constructed socially, historically, and spatially.

Third, regime theory's conception of capitalism and, more generally, capitalist society, leads it to draw a sharp distinction between state and market. Perhaps some would invoke a third entity, civil society, to lump together community organizations, churches, nonprofits, and other empirical entities involved in urban politics. But this underestimates how plastic capitalism is. If a capitalist in New England finds adequate supplies of suitably trained labor at private trade schools, why should he (and it usually is a he) be involved in local politics to initiate a state-run apprenticeship program? It the capitalist has trouble recruiting labor, he can automate, move, subcontract, and so on. The choice to solve the problem by working with local government is *contingent* and *reflects* the local regime as much as it constitutes it.

Similarly, local government has many alternatives within the scope of regime theory's basic assertions of business' key role in local policy and that successful governing regimes use policy to maintain the governing coalition. Yet local policy decisions are not just matters of choosing among alternatives within these constraints; other factors pressure, constrain, and enable policy. For instance, we would not expect urban regimes today to build barge canals as a development strategy, as was common in the early 19th century. Technology, ideology, local industrial history, the locality's place in the spatial division of labor, and so on all enter the equation, and regime theory's conceptual apparatus alone offers little to decipher these forces.

Regime theory has been overly concerned with real estate and other business groups directly involved in urban development, but it has downplayed numerous "nondecisions" involving manufacturers and other sectors of capital only indirectly affected by local government projects. Even with real estate, regime theory's conception of capital has been one-dimensional. Is it hard to understand why developers became actively involved in Atlanta, which grew from a population of under 700,000 in 1950 to about 2.5 million in 1985, whereas real estate seemed to be much less dominant in Dayton during a period that Dayton lost 25% of its blue-collar jobs as part of an atrophying auto-tire complex (Clarke, 1987; Storper & Walker, 1989). To be sure, one might argue that regime politics *caused* Atlanta's growth and Dayton's decline, but then this needs to be demonstrated. On the other hand, if the respective regimes did not make this difference, how does "politics matter"?

This brings up the second, and in my view more fruitful, way to integrate regime theory into regulation theory. Instead of focusing on politics and policy outcomes, regime theory could help us understand local economic dynamics by helping us understand how the local state affects the local SSR. This implies broadening regime theory beyond "urban development" to encompass everything bound up in the dynamic reproduction of the SSR: the environment, education, crime, housing, and so on. Besides overt policy, we must also examine more subtle aspects of the local state, such as how it naturalizes and legitimizes certain discourses over others, its history of success and corruption, and so forth.[13] Finally, we must put the state in its place by recognizing it is not the only game in town. Other regulatory mechanisms, including other formal institutions and informal things like local norms or geographic morphologies, can substitute for the local state or limit its role in regulating the local SSR.

In sum, regime theory can add to our understanding of the local state as a site of *contingent* outcomes. However, these outcomes in and of themselves hold relatively little interest. They become more interesting when they play an important role in the dynamic processes that reproduce and change socioeconomic space. To decipher this role, one needs to mesh regime theory with a systematic and developed theoretical understanding of social and economic change in space and time, which regulation theory aspires to provide. Yet regulation theory, as it has been developed thus far, lacks a conceptual apparatus capable of accounting for the dynamic spatial diversity-in-unity that characterize accumulation regimes. Perhaps the more detailed concepts discussed here can provide a suitable bridge between regime and regulation theory that not only integrates them but also transforms them in the process.

# Notes

1. On these points, see the essays by Cox, Goodwin and Painter, Jonas, and Lauria in this volume. Also see Peck and Tickell (1992) and Tickell and Peck (1992).

2. This does not deny that regime theory helps us understand local politics or see that local governments do have some autonomy and can have some impact on capitalist economies.

3. Here "process" captures several ontological properties: integration of social relations and flows, distinct patterns, inherent dynamics, and ongoing reproduction and change.

4. "Materials" pertains to all physical entities, including, for example, energy and radio signals.

5. As is common practice, I characterize commodities, prices, values, and finance as flowing in "circuits." Because flows of materials, personnel, information, and property rights are not necessarily closed or circular, the less restrictive term *process* describes them more accurately. Note, however, the strong family resemblance between capital's traditional circuits and these other "circuits" or processes.

6. Obvious ties include gender, race, and so forth—elements that both regime and regulation theory have tended to downplay. Alas, space limitations preclude discussing these ties here.

7. We might include symbolic and discursive resources under "information." Yet insofar as all human activity involves thought, the other processes include such resources too. Symbolic and discursive resources are integral to the social construction of norms and practices and therefore might be thought of as distinct from the more overt and specific flow of information.

8. See Walker (1988) and Polanyi (1944) for similar discussions of coordinating mechanisms. This typology is just the tip of the iceberg. We also need to understand how the processes are reproduced, their dynamics, contradictions, and so on. Typically, a regime of regulation involves multiple means of regulation in combination.

9. "Transaction cost" theory explains the boundary of the firm in terms of efficiency: When market transactions are too costly, integration will result. This not only leads to an overly simple dichotomy—firm versus market—it also overlooks the need for control when a subordinate's goals and interests conflict with a superior's. The superior must then find a way to *force* compliance, and this can be difficult outside the firm.

10. Transaction cost economics frequently makes the erroneous assumption that other means of regulation do not entail transaction costs.

11. Here I italicize terms from the previous discussion to highlight its relevance.

12. Of course, this does not mean this model of postfordism can resolve fordism's crisis or reestablish a virtuous circle of stable growth. Nor do I mean to imply that competition does not drive out *some* industrial practices. Also, one might argue that earlier periods integrated diverse local systems into a single world system. Nonetheless, I believe the current period's logic of integration and the specific local systems being integrated are distinct. Rather than consider regions that do not conform to a mythical model of universal regional development as retrograde, we need to understand their place in the world system and how they survive and change.

13. See Horan (1991) for some potentially fruitful thoughts along these lines.

# 4

# A Neo-Gramscian Approach to the Regulation of Urban Regimes

*Accumulation Strategies,*
*Hegemonic Projects,*
*and Governance*

BOB JESSOP

The perspective on urban regimes presented here is based on a neo-Gramscian reading of the regulation approach as well as on a classically Gramscian account of the necessary reciprocal relations between state and civil society. In particular I argue that urban regimes can be fruitfully analyzed as strategically selective combinations of political society and civil society, of government *and governance*, of "hegemony armored by coercion." I also argue that such regimes may be linked to the formation of a local hegemonic bloc (or power bloc) and a historic bloc (or accumulation regime and its mode of regulation). To support these proposals I present some of Gramsci's ideas about the relations between

AUTHOR'S NOTE: This chapter arises from an Economic and Social Research Council (ESRC) research program on local governance, Grant No. L311253032. It has benefited from comments by my research officer, Gordon MacLeod, other participants in the ESRC program, and the editor and contributors to the present volume. The usual disclaimers apply.

the economic base and its superstructure as well as his key concepts for analyzing the state and state power. This enables me to show marked similarities between a Gramscian account of the state and regulationist views on the capital relation. In the light of these analogies I outline eight key lessons for an exploration of urban regimes inspired by a neo-Gramscian regulation approach and, drawing on more recent work in the regulation approach, I offer some reflections on current changes in urban regimes. My argument is shaped by European experience, but I would also claim that the neo-Gramscian approach can be fruitfully applied elsewhere.

## Theoretical Perspectives on Urban Regimes

I now present three theoretical perspectives for an alternative research agenda on urban regimes. After briefly presenting some Gramscian insights into politics and key ideas from the French regulation approach, I consider Gramsci's ideas about the ethico-political dimension of economic regimes. This enables me to propose some parallels between Gramscian and regulationist approaches to political economy. I also comment on the relevance of governance theory to both approaches.

### Gramsci and Integral Politics

Gramsci (1971) analyzed the state "in its inclusive sense." He defined it as "the entire complex of practical and theoretical activities with which the ruling class not only justifies and maintains its dominance but manages to win the active consent of those over whom it rules" (p. 244). This approach is linked to his equation of the state with "political society + civil society" and his claim that state power in the West rests on "hegemony armored by coercion" (pp. 261-263). Gramsci did not examine the constitutional and institutional features of government, its formal decision-making procedures, or its general policies (the state in its narrow sense, so to speak); instead, he explored how political, intellectual, and moral leadership was mediated through a complex ensemble of institutions, organizations, and forces operating within, oriented toward, or located at a distance from the juridico-political state apparatus. This suggests that the political sphere can be seen as the domain where attempts are made to (re)define a "collective will" for an imagined political community[1] and to (re)articulate various mechanisms and practices of government and governance in pursuit of projects deemed to serve it. Although his prison writings dealt mainly with national-popular politics in national states (especially Italy and France), nothing excludes its application to urban politics. This can be seen from Gramsci's notes on communal politics in medieval Italy as well as incidental remarks on contemporary cities such as Turin, Rome, and Naples. More generally, one could argue that his approach is actually highly relevant to local politics because it downplays the importance of sovereign states with their monopoly of coercion and allows more weight to other apparatuses, organizations, and practices involved in exercising political power.

## The Regulation Approach and Integral Economics

Regulationists study the economy in an inclusive sense. In a manner reminiscent of Gramsci's expanded treatment of the state (as *lo stato integrale* or the integral state), they investigate what one could likewise call *l'economia integrale* (or the integral economy). Expressed in other words, regulation theorists examine the historically contingent ensembles of complementary *economic* and *extra-economic* mechanisms and practices that enable capital accumulation to occur in a relatively stable way over long periods despite the fundamental contradictions and conflicts generated by the capital relation itself (cf. Aglietta, 1979; Boyer, 1990; Lipietz, 1987). In particular, although far from neglecting the essentially anarchic role of exchange relations in mediating capitalist reproduction, regulationists tend to emphasize the complementary role of other mechanisms (institutions, norms, conventions, networks, procedures, and modes of calculation) in structuring, facilitating, and guiding (in short, regulating) capital accumulation. Regulationists argue that relatively stable capitalist expansion over any extended time period depends not only on economic institutions and practices but also on crucial extra-economic conditions that cannot be taken for granted. In this context, regulation theorists must go well beyond any narrow concern with production functions, economizing behavior, and pure market forces to study the wide range of institutional factors and social forces that are directly or indirectly or both involved in capital accumulation.

In the pioneer analysis in this approach, Aglietta (1979) studied how regulation "creates new forms that are both economic and noneconomic, that are organized in structures and themselves reproduce a determinant structure, the mode of production" (pp. 13, 16). Thus, he was initially interested in what one might call both the *economic* and the *social* modes of economic regulation. The economic mode would refer to the key role of economic exchange and market forces in the self-organization of capitalism, and the social mode would refer to the role of the extra-economic in organizing economic activities. Among other factors frequently mentioned by regulationists in this regard are the legal and social regulation of the wage relation, the articulation of financial and industrial capital, forms of corporate organization, modes of economic calculation, the role of the state, education and training, and international regimes. Interestingly, since Aglietta's initial work, many regulationists have increasingly focused on the social mode of economic regulation to the neglect of the economic. This seems to move them even closer, albeit implicitly, to a Gramscian perspective on the expanded reproduction-regulation of the capital relation. This impression is reinforced by references, this time quite explicit, by some key regulation theorists to the need to strengthen their account of the state (a topic they regard as undertheorized within the French regulation approach) with Gramscian or neo-Gramscian ideas (e.g., Aglietta, 1979; Häusler & Hirsch, 1987; Jenson, 1990; Lipietz, 1987, 1994; Lordon, 1995; Noël, 1988; for further details, see Jessop, 1990b, pp. 311-319).

### Once More on Gramsci

To reinforce these points I now return to Gramsci's approach to economics and politics. Whereas regulationists' main concern has been to offer a comprehensive economic analysis of the socially embedded, socially regularized nature of the capital relation to better understand its relative stability as well as its crisis tendencies and transformation, Gramsci's chief concern was to develop an autonomous Marxist science of politics in capitalist societies and thereby establish the most likely conditions under which revolutionary forces might eventually replace capitalism. These apparently contrasting theoretical (as opposed to practical) interests are by no means inconsistent. Indeed, in discussing different modes of inquiry and knowledge, Gramsci (1971) observed that

> in economics the unitary centre [of analysis] is value, alias the relationship between the worker and the industrial productive forces . . . In philosophy [it is] praxis, that is, the relationship between human will (superstructure) and economic structure. In politics [it is] the relationship between the State and civil society, that is, the intervention of the State (centralised will) to educate the educator, the social environment in general. (pp. 402-403; cf. remarks on the difference between economic and political interests and calculation, p. 140)

Although Gramsci spent most of his prison years working on philosophy and politics, he also offered various comments that anticipated the regulation approach. This is most obvious, of course, in his pioneering remarks on Americanism and fordism and the prospects of introducing them into a Europe with a very different history and civilization (Gramsci, 1971, pp. 277-318; Gramsci, 1995, pp. 256-257). But he also offered some important methodological remarks on economic analysis. In particular we may note how he redefined Ricardo's concept of "determined market" [*mercato determinato*][2] as "equivalent to [a] determined relation of social forces in a determined structure of the productive apparatus—this relationship being guaranteed (that is, rendered permanent) by a determined political, moral, and juridical superstructure" (Gramsci, 1971, p. 410). Gramsci continued that, whereas classical economists had treated determined market as an arbitrary abstraction and reified its various elements and laws as "eternal" and "natural," Marxist political economy began from the historical character of "determined market" and its social "automatism" and studied these phenomena in terms of "the *ensemble* of the concrete economic activities of a determined social form" (Gramsci, 1971, p. 400n, 411; cf. Gramsci 1995, pp. 172, 427). Thus, according to Gramsci, economic laws (necessities, "automatism") should be understood as tendencies, located historically, grounded in specific material conditions, and linked to the formation of a specific type of "homo oeconomicus," reflected in turn in popular beliefs and a certain level of culture (Gramsci, 1971, pp. 279-318, 412, 400n, 413; Gramsci, 1995, p. 167). Economic laws or "regularities" (*regolarità*) are secured in so far as the actions of one or more strata of intellectuals give the dominant class a certain homogeneity and an awareness of its own function in the social *and political* as well as the economic fields (Gramsci, 1971, pp. 410-414, emphasis added). For it is essential that entrepreneurs organize "the general system of relationships external to the business itself" (p. 6). In this context Gramsci (1971) noted that the "conquest of power and achievement of a new productive world are inseparable, and that propaganda for one of them is also propaganda for the other, and that in reality it is solely in this coincidence that the unity of the dominant class—at once economic and political—resides" (p. 116).

These ideas can be developed by considering two further insights. One is Gramsci's basic analytical distinction between *historical bloc* and *power bloc*. The first term in this conceptual couplet has important implications for the regulation approach; the second has implications for work on growth coalitions. The other insight is expressed in Gramsci's account of the "decisive economic nucleus" necessary for hegemonic projects to be successful in the long run. This account also has important implications for urban regimes, especially the social and economic bases of different growth coalitions therein.

Gramsci (1971) employs the notion of historical bloc to solve the Marxian problem of the reciprocal relationship between the economic base and its politico-ideological superstructure. He addresses how "the complex, contradictory and discordant ensemble of the superstructures is the reflection of the ensemble of the social relations of production." This issue is typically analyzed dialectically in terms of how the historical bloc reflects "the necessary reciproc-ity between structure and superstructure" (p. 366). This reciprocity is realized, according to Gramsci, through specific intellectual, moral, and political prac-tices that translate narrow sectoral, professional, or local (in short, in Gramscian terms, "economic-corporate") interests into broader "ethico-political" ones. Only thus does the economic structure cease to be an external, constraining force and become a source of initiative and subjective freedom (pp. 366-367). In this sense the ethico-political not only co-constitutes economic structures but also provides them with their rationale and legitimacy. Gramsci adds that analyzing the historical bloc can show how "material forces are the content and ideologies are the form, though this distinction between form and content has purely didactic value" (p. 377). Although Gramsci does not introduce regulationist concepts such as "industrial paradigms, models of development" (Lipietz, 1987), "accumulation strategies" (Jessop, 1983), or "societal para-digms" (Jenson, 1990, 1993) in this context, such concepts certainly help to illuminate the ethico-political moment of the historical bloc. For these concepts bring out the importance of values, norms, vision, discourses, linguistic forms, popular beliefs, and so on, in shaping the realization of specific productive forces and relations of production.

In this spirit, a historical bloc can be defined as a historically constituted and socially reproduced correspondence between the economic base and the polit-ico-ideological superstructures of a social formation. Stripped of its historical materialist "base-superstructure" jargon, this concept is easily redefined in Parisian regulationist terms. Thus, a historical bloc could be understood as the complex, contradictory, and discordant unity of an accumulation regime (or mode of growth) and its mode of economic regulation.[3] The dialectical relation-ship between form and content could then be seen to develop through what one could interpret as the co-constitution of the accumulation regime as an object of regulation in and through its co-evolution with a corresponding mode of regu-lation (cf. Jessop, 1990b, p. 310; see also Painter, this volume). Or to paraphrase Gramsci's own comments on the state and state power, one could say that the economy in its inclusive sense comprises an "accumulation regime + mode of regulation" and that accumulation occurs through "self-valorization of capital in and through regulation."

The concept of hegemonic bloc was introduced in Gramsci's discussion of class alliances or national-popular forces mobilized in support of a particular hegemonic project. It refers to the *historical unity,* not of structures (as in the case of the historical bloc) but of *social forces* (which Gramsci analyzed as the ruling classes, supporting classes, mass movements, and intellectuals). An hegemonic bloc is a durable alliance of class forces organized by a class (or class fraction) that has proved itself capable of exercising political, intellectual, and moral leadership over the dominant classes and the popular masses alike. Thus, Gramsci notes that "the historical unity of the ruling classes, . . . results from the organic relations between State or political society and 'civil society' " (1971, p. 52). Although this argument applies principally to the national state, it can also be used in studying supra- and subnational regimes (see, for example, van der Pijl, 1982, on the transatlantic ruling class in Atlantic fordism; Dulong, 1978, on the local state and hegemony in French regions and communes).

Significantly, Gramsci (1971) recognizes several degrees and forms of political rule—not all of them fully hegemonic. They range from an inclusive hegemony that secures the active consent of the majority of all classes; through more limited forms of hegemony based on selective incorporation of subordinate groups (or at least their leaders) and limited, piecemeal material ("economic-corporate") concessions; to a resort, in exceptional cases, to generalized coercion (pp. 105-106). Gramsci remarks, for example, that the dominant economic class in Italy's medieval communes was unable to create its own category of intellectuals and so failed to build a solid hegemony. The communes had a more confederal, "syndicalist" nature: Rather than having a hegemonic bloc, they rested on a mechanical bloc of social groups, often of different races, with some subaltern groups having parastatal institutions of their own and enjoying considerable autonomy within broad limits set by coercive police powers (pp. 54n, 56n). Elsewhere Gramsci criticizes urban politics in nonindustrial cities, such as Naples, which serve primarily as unproductive centers for regional government and the consumption of parasitic classes and strata; he also notes that these dominant intellectual strata are more likely to be "pettifogging lawyers" than the technocrats who predominate in northern industrial cities (pp. 90-94, 98-100).

The final insight to be explored here is Gramsci's (1971) observation that "though hegemony is ethical-political, it must also be economic, must necessarily be based on the decisive function exercised by the leading group in the decisive nucleus of economic activity" (p. 161). This claim would seem open to several interpretations. It could be taken to imply that only the bourgeoisie (or its dominant fraction) could really exercise hegemony once capitalism has emerged, and that only the proletariat is in a position to develop an organic counterhegemonic project. But given the mediating role of organic intellectuals (who need not themselves be recruited from the two fundamental or decisive classes), this interpretation is far too class reductionist and instrumentalist. Thus, Gramsci's claim could be better read in an (integral) economic manner as implying that the essential function of a hegemonic project is to secure the (integral) economic base of the dominant mode of growth, and that it does this through the direct, active conforming of all social relations to the economic (and extra-economic) needs of the latter. Thus, Gramsci argues that

> every State is ethical in as much as one of its most important functions is to raise the great mass of the population to a particular cultural and moral level, a level (or type) which corresponds to the needs of the productive forces for development, and hence to the interests of the ruling classes. (p. 258)

A third interpretation can be framed in integral political terms. Here Gramsci's (1971) comment could be taken to mean that all feasible organic hegemonic projects needed to respect (or take account of) "economic determination in the last instance." Gramsci argues the economy is nothing but "the mainspring of history in the last analysis" (p. 162). Only by examining forms of consciousness and methods of knowledge can one decipher the necessarily indirect impact and repercussions of economics within the wider society (see pp. 162, 164, 167, 365). Thus, "an analysis of the balance of forces—at all levels—can only culminate in the sphere of hegemony and ethico-political relations" (p. 167). In this sense, political forces have a vested interest in securing the productive potential of the economic base that both generates political resources and defines the scope for making material concessions. Wealth must first be produced before it can be distributed. From an integral political viewpoint, this does not mean that economic growth is invariably accorded the highest political priority—even when such growth is understood integrally. It implies only that political agents must be concerned with the economic conditions of juridico-political or politico-military power and be sensitive to the political effects of economic developments. Thus, although certain economic-corporate interests of (fractions of) the bourgeoisie can be sacrificed, the essential foundations of capitalism must be respected. In addition to hegemony directly and explicitly based on an accumulation strategy, therefore, hegemony could also establish other priorities provided that the core conditions for capital accumulation are not thereby irrevocably undermined.

## Governance Theory

The final term to be introduced here is *governance*. This is increasingly popular both as an umbrella term for all forms of coordination of social relations (the "conduct of conduct") and as a more specific (but still very general) term for forms of coordination that involve neither market forces nor formal hierarchy. In this latter sense, economic analysts often refer to novel forms of economic coordination such as relational contracting, "organized markets" in group enterprises, clans, networks, business or trade associations, strategic alliances, and various international regimes. Likewise, in political science, disquiet has grown with a rigid public-private distinction in state-centered analyses of politics and its associated top-down account of the exercise of state power. This is reflected in concern with the role of various forms of political coordination that not only span the conventional public-private divide but also involve "tangled hierarchies," parallel power networks, or other forms of complex interdependence across different tiers of government or different functional domains. A general definition encompassing all of these forms is that governance refers to the "self-organization of inter-organizational relations" (cf. Jessop, 1995b).

Governance is significant both for neo-Gramscian political analysis and for the regulation approach. It is relevant to the "micro-physics" of power, that is, the channels through which diverse state projects and accumulation strategies are pursued and, indeed, modified during their implementation. Because state power is inevitably realized through its projection into the wider society and its coordination with other forms of power, one must look beyond formal government institutions to a wide range of governance mechanisms and practices. Likewise, governance is relevant to the day-to-day practices in and through which the various structural forms of regulation are instantiated and reproduced. Regulationists typically define these forms in institutional terms (hence, the frequent charge against them of structuralism) to the neglect of specific practices and emerging conflicts. One can compensate for this bias by examining economic governance and how expectations are stabilized within particular structural contexts and behavior is regularized through conventions, compromise, and the exercise of power (for some initial regulationist work in this area, see Benko & Lipietz, 1994; Boyer & Hollingsworth, 1995; Lipietz, 1993; for a review, see Jessop, 1995b). Early studies of urban regimes also offer important insights into the nature of governance and more recent work makes explicit use of the concept in posing research issues (for a recent review, see Stoker, 1995). In short, urban governance seems an important area for assessing the potential of a neo-Gramscian regulation approach.

# On Analyzing Urban Regimes

That urban regimes could be a fruitful test bed in this regard is indicated by Stoker's (1995) comment that regime theory emphasizes "the interdependence of governmental and non-governmental forces in meeting economic and social challenges" (p. 54; cf. Stone, 1993). Moreover, given that a neo-Gramscian regulationist approach is concerned with the economy in its inclusive sense, it seems even more suited to dealing with the role of government and governance in addressing contemporary economic challenges. In this spirit I now suggest eight lessons for studying local economic governance. These largely derive from the preceding theoretical review (supplemented on occasion with arguments from my earlier work) and are meant to serve as analytical guidelines rather than as research hypotheses. Before presenting these guidelines, however, two general cautions are needed. The following lessons are concerned with economic governance, but by no means do all urban regimes give this issue the highest priority. Moreover, given the rigorous conditions they establish for success in this regard, these guidelines are best understood as a negative heuristic, for they are more likely to serve in most case studies to identify sources of governance failure than success.

The first lesson concerns objects of governance. Those regulation theorists opposed to functionalist arguments regard modes of regulation as being constitutive of the objects they regulate: Objects and modes of regulation co-evolve in a structurally coupled (and often, indeed, strategically coordinated) manner (on this, see Jessop, 1990b; Painter, this volume). Thus, one should study how the local economy comes to be constituted as an object of economic *and* extra-economic regulation. This involves examining two interlinked distinctions: (a) the local economy versus its supralocal economic environment and (b) the local economy versus its extra-economic local environment (community, the political system, welfare state, education system, religious institutions, and so on). The first distinction is premised on the idea that, whatever the vagaries and contingencies of economic development on a global scale, the possibility exists to endogenize and control at least some conditions bearing on local economic development. At stake here is how the boundaries of the local economy are discursively constructed and materialized. The second distinction refers to means-ends relations in developing local strategies from an integral economic perspective and concerns the range of activities that must be co-coordinated to realize a given economic development strategy.

The second lesson derives from the regulation approach and work on the governance of complexity. Both the supralocal economic environment and the extra-economic local environment are more complex than local economic actors can understand (especially in real time), and both will always involve a more complex web of causality than they could ever control (because adequate control would require that local economic actors command diverse means of influencing the interaction of causal mechanisms over time and space corresponding to the complexity of those mechanisms). Moreover, as economic and economically relevant activities increasingly extend over larger spatial scales, it gets harder to demarcate a relatively autonomous economic space at less than global scale. Thus, we must direct attention to the role of the spatial imaginary and economic narratives or discourses in demarcating a local economic space with an imagined community of economic interests from the seamless web of a changing global-regional-national-local nexus. There is no reason, of course, why such a subset of economic relations should coincide with a given political space (whether defined territorially or organizationally). Nor is there any reason a priori why economic rhythms should coincide with cycles or rhythms related to localized forms of government and governance or with their over determination by outside forces. Thus, we must also consider the specific practices, if any, that tend to transform this into a real space amenable to regulation or governance practices that are concerned to realize these common interests over a given time horizon. According to Gramsci's views on the reciprocal necessity of base and super-structure, a key role in both respects would fall here to intellectual forces (broadly understood) involved in elaborating the "ethico-political" aspects of the relevant historical bloc.

Lesson three introduces the neo-Gramscian concept of accumulation strategy to build on this double demarcation of a manageable economic space and its extra-economic conditions. Struggles over the economic *and social* modes of economic regulation play a key role in shaping and unifying different suprana-tional, national, regional, and local modes of growth. Because the different structural forms of the capitalist economy (the commodity, money, wage, price, tax, and company forms) are generic features of all capitalist economic relations and are unified only as modes of expression of generalized commodity produc-tion, any substantive unity that characterizes a given capitalist regime in a given economic space must be rooted elsewhere. One such source is accumulation strategies. These define a specific economic growth model for a given economic space and its various extra-economic preconditions and they outline a general strategy appropriate to its realization. As noted earlier, organic intellectuals linked to the dominant class play a key role here. It must be admitted that, regarding fordism, Gramsci (1971) claimed organic intellectuals' role would be greater in interwar Europe, which had a more complex class and social structure than the United States; in the latter, the emerging hegemony of fordism was more securely rooted in the factory and had less need of professional political and intellectual intermediaries (p. 285). Whatever the merits of his argument for the emergence and consolidation of fordism on both sides of the Atlantic, it is evident that major political, intellectual, and moral struggles have occurred in shaping the emerging postfordist modes of regulation with their new, more flexible *homo oeconomicus,* new norms of production and consumption, new discourses and societal paradigms, new structural forms and institutional supports, and new modes of government and governance. Accumulation strategies can be defined for different spatial scales from international regimes through supranational blocs to national and regional economies and thence to the local. Although this concept has generally been applied to the national level (itself a reflection of the dominance of national economies and national states in the fordist era), it is also relevant to the regional and local level. Indeed the crisis of fordism has made it even more relevant (see later).

Lesson four concerns the need to examine the relationship between local accumulation strategies and prevailing hegemonic projects. Because economies (even in their inclusive sense) are always embedded in a wider ethico-political context, the stability of the ethico-political context also merits attention. A key role can be played here by hegemonic projects that help secure the relative unity of diverse social forces. A hegemonic project achieves this by resolving the abstract problem of conflicts between particular interests and the general interest. This project mobilizes support behind a concrete program of action that asserts a contingent general interest in the pursuit of objectives that explicitly or implicitly advance the long-term interests of the hegemonic class (fraction) and thereby privileges particular economic-corporate interests compatible with this program while derogating the pursuit of other particular interests that are inconsistent with the project. Moreover, although the hegemonic project serves the long-run interests of the dominant class (or class fraction), this class will typically sacrifice certain economic-corporate interests in the short term to help legitimate its overall hegemonic project. For Gramsci, hegemony was generally realized at the national-popular level and expressed in the organization of national states; more recent studies have explored international hegemony in Gramscian terms. If, however, one principal feature of contemporary capitalism is the "hollowing out" of national states and the resurgence of regions and cities, it is important to consider how far hegemony can also be relocated (perhaps once again) at the subnational level. This is particularly likely in more federal or decentralized regimes, but there are also clear signs of a resurgence of municipal governance in unitary states. Nonetheless we should note that, just like national hegemony, local hegemonies can also vary in their relative inclusiveness, the balance between active consent, fraud-corruption, and coercion, the relative weight of government as opposed to governance. It would be interesting to explore differences here between urban regimes in, for example, export-led flexible industrial districts in boom regions and property-led urban regeneration in crisis-prone fordist cities.

The fifth lesson derives from state-theoretical (and, more broadly, strategic-relational) arguments that institutional ensembles involve specific forms of strategic selectivity. A major problem with Gramsci's analysis of hegemony was its emphasis on the changing balance of social forces at the expense of the underlying balance of power inscribed within specific structures. At most, he discussed metaphors such as war of position and war of manoeuvre. But it is important to analyze both how far, and the manner in which, institutions and apparatuses are strategically selective, that is, involve a structurally inscribed mobilization of strategic bias.[4] Particular forms of economic and political system privilege some strategies over others, access by some forces over others, some interests over others, some spatial scales of action over others, some time horizons over others, some coalition possibilities over others (cf. on the state, Jessop, 1990b, pp. 260ff.). Structural constraints always operate selectively: They are not absolute and unconditional but always temporally, spatially, agency, and strategy specific. This has implications both for general struggles over the economic and extra-economic regularization of capitalist economies and specific struggles involved in securing the hegemony of a specific accumulation strategy. Although the idea of strategic selectivity (and its precursor, structural selectivity) was initially developed in analyses of the state, it has obvious implications for research into modes of growth. Here it refers to the differential impact of the core structural (including spatiotemporal) features of a labor process, an accumulation regime, or a mode of regulation on the relative capacity of particular forces organized in particular ways to successfully pursue a specific economic strategy over a given time horizon and economic space, acting alone or in combination with other forces and in the face of competition, rivalry, or opposition from yet other forces.

Combining this approach with work on the strategic selectivity of governance regimes should offer powerful tools for urban regime analysis. Not all economic and political forces derive the same advantages from specific modes of growth or governance. Mapping such asymmetries has an important role in defining the nature of urban regimes—especially as the strategic selectivity of local institutions affects their long-run stability. Indeed, the durability of urban regimes depends not only on the overall coherence and economic feasibility of the strategies they promote but also on strategic capacities rooted in local institutional structures and organizations. This said, of course, another implication of the strategic-relational approach is that agents are reflexive, capable of reformulating within limits their own identities and interests, and able to engage in strategic calculation about their current situation (see Jessop, 1982, 1996b). This opens the possibility of strategic action to transform the strategic selectivity of extant regimes. For example, much of the Thatcherite attack on local government autonomy in Britain and the associated transfer of local authority functions was due to a concern to redefine access to local power and promote "enterprise culture."

The sixth lesson is more clearly neo-Gramscian and concerns the scope of such power structures. It is important to examine how urban regimes operate through a strategically selective combination of political society *and civil society,* government *and governance,* "parties" and partnerships. In this way one could show how some urban regimes can be linked to the formation of a local hegemonic bloc (or power bloc) and its associated historical bloc. Nonetheless one must recall that Gramsci (1971) himself allowed for a wide range of power structures and modes of exercising rule—not all of which involve an inclusive form of hegemony based on active consent. He also emphasized that politics could not be read off mechanically from the economic base and noted that many features of politics (especially in the short term) are due to political miscalculation, the impact of specific political conjunctures, or organizational necessities of different kinds that have little, if any, direct connection to the economic base (pp. 408-409). Conversely, he emphasized that viable hegemonic projects (and, one might add, accumulation strategies) must have some organic connection to the dominant mode of growth. They cannot simply be "arbitrary, rationalistic, and willed" but must have some prospects of forming and consolidating a specific historical bloc (pp. 376-377).

The seventh lesson is that, whatever specific structural forms and political projects sustain or, at least, privilege, the ruling bloc in an urban regime, a need also exists for an adequate repertoire of governance mechanisms and practices to ensure the ruling bloc's continued vitality in the face of a turbulent environment and emergent conflicts that threaten the unstable equilibrium of compromise on which the bloc is based. Governance failure will have a serious impact on the relative stability of specific urban regimes and the success of different local economic strategies.

The eighth lesson is cautionary. It would be a gross mistake to assume that a local mode of growth, a local mode of regulation, or an urban regime can exist in isolation from its environment. Although this comment may seem unnecessary regarding local modes of growth, discussions of modes of regulation tend to neglect how far both the social and economic aspects of regulation are embedded in tangled hierarchies for any given spatialized object of regulation. Just as the postwar Atlantic fordist mode of national economic growth had international and regional/local supports, so does the mode of growth of regional and local economies depend on more encompassing economic complementarities, structural forms, modes of governance, and so forth. Strictly speaking, it would be more appropriate, if somewhat convoluted, to talk about plurispatial, multitemporal, and polycontextual modes of regulating local economies and their relative integration into more encompassing economic spaces. Analogous problems have already been noted in criticisms of urban growth coalition theories (see Harding, 1995).

## Local Economies and Economic Strategies

In depicting a local accumulation regime we encounter a definitional problem that not only is present for observers but also affects local participants. This is how to demarcate a local economy and its economic and extra-economic conditions of existence and, on this basis, to formulate a local accumulation strategy concerned with local economic development. Such strategies can be defined for various economic units (both territorial and functional), but my concern here is with possible features of local accumulation strategies associated with specific urban regimes. The variable geometries of economic and political boundaries pose major problems concerning whether local political forces have the juridico-political capacities to manage or govern the local economy. This is often noted for the United States but also occurs elsewhere. Any solution depends as much on the spatial imaginary and the links between state and civil society, however, as on formal territorial demarcations and the reallocation of formal legal and political powers. Once one adopts an integral economic and integral political approach to local economic development, one can see how local economies and local regimes might be organized across borders. There is clearly a key role here for local growth coalitions (broadly understood to comprise the major forces mobilized behind the dominant local accumulation strategy rather than limited property development coalitions) in shaping the conditions for local economic performance.

The choice of spatial scale where local economic development should be pursued is inherently strategic and contingent on various political, economic, and social specificities of a particular urban and regional context at a particular moment. The temporal and spatial are not separable here. The choice of time horizon will in part dictate the appropriate spatial scale at which development is sought. In turn, the choice of spatial scale will in part determine the time horizon within which local economic growth can be anticipated. Thus, the discursive constitution of the boundaries and nature of the (local) economy affects the temporal dimension of strategy making as well as its spatial scale. This is quite explicit in many economic strategy documents—with powerful players seeking to shape both the spatial and temporal horizons to which economic and political decisions are oriented so that the economic and political benefits are optimized (on the case of the East Thames Corridor in Britain, see, for example, Jessop, 1996a). Hence, the ability to match spatial scale and time horizon can be a crucial factor shaping the success or failure of local economic development strategies associated with urban regimes. When space and time horizons are articulated more or less successfully, economic development will occur within "time-space envelopes" (cf. Massey, 1994, p. 225; Sum, 1995).

Regarding the supralocal economic environment and the extra-economic local environment, the attempted governance of complexity involved in local accumulation strategies requires key players to undertake two interrelated tasks. These are, first, to model the factors relevant to local economic development based on the analytical distinction between the local economy and its two above-named environments, and second, to develop the requisite variety in policy instruments or resources to be deployed in the pursuit of local accumulation strategies. This puts considerable demands on the monitoring and self-reflexive capacities of local growth coalitions and suggests the importance of their own organizational learning capacities as well as those of the local or regional economy as a whole. The greater the capacities of a specific group or network to learn, the greater the chances of its becoming hegemonic in defining the local accumulation strategy, and in addition, that the latter will be organic rather than "arbitrary, rationalistic, and willed." In both respects economic hegemony also requires acceptance of the strategy by other key players whose cooperation is needed to deliver the extra-economic conditions to realize an accumulation strategy.

A final point to note is the extent to which local economic and political forces can draw on wider sources of knowledge about the economic and extra-economic conditions that bear on the competitiveness of local economies. For stable modes of local economic growth typically involve building a structured complementarily (or coherence) between the local economy and one or more of its encompassing regional, national, and supranational accumulation regimes. Capitalism is always characterized by uneven development and tendencies toward polarization, so the success of some economic spaces (and the success of the spaces whose growth dynamic is complemented by their own) will inevitably be associated with the marginalization of other economic spaces. This is seen in the changing hierarchy of economic spaces as capitalist growth dynamics are affected by the relative exhaustion of some accumulation strategies and modes of growth or the dynamic potential of innovations in materials, processes, products, organization, or markets. This in turn means that different urban growth coalitions should orient local accumulation strategies to an assessment of the position of their local economic space in the urban hierarchy and international division of labor. This explains the wide range of alternative strategies in different localities, highlighting the need for economic development initiatives that are sensitive to the specificities of particular local economies (see Barlow, 1995; Hay & Jessop, 1995; Krätke, 1995).

These considerations make it important that strategically reflexive actors on the local scene try to choose appropriate spatial and temporal horizons of action as well as appropriate strategies and tactics to improve their chances of realizing their aims and objectives. Yet any attempt to isolate spatiotemporally a set of social relations from the complex and continuous web of causal connections is inherently fragile and bound to produce unintended consequences. These consequences will be harder to deal with and learn from to the extent that the environment is more turbulent or the system more complex or both. Moreover, although all actors routinely monitor the effects of their actions, such turbulence and complexity obviously constrain their ability to engage in strategic (including organizational) learning.

## Changing Urban Regimes

We can now attempt a reinterpretation of the changed economic agenda of economic partnerships in contemporary western capitalism. It was always one-sided to suggest that local growth coalitions were oriented primarily to property development. For during fordism, as now, strategies pursued by local authorities and agencies of local governance were numerous, spatially and temporally specific, and divergent. But it is fair to say that the local coalitions were oriented to specific models of development that generally complemented the dominance of Atlantic fordism and its distinctive forms of uneven development. Thus, local states under fordism typically provided a local infrastructure to support fordist mass production, promoted collective consumption, implemented local welfare state policies, and in some cases (especially as the crisis unfolded), engaged in competitive subsidies to attract new jobs or prevent the loss of established jobs. Although local economic conditions clearly shaped how individual local governments saw their respective *economic* roles, there was an almost universal commitment to the Keynesian welfare *social* policy role. These generalizations apply most strongly, of course, to trends in Europe. Indeed, as Gramsci himself noted, hegemony was more rooted in the factory in the United States; among other things, this meant that Keynesian welfare was supplied in part through company- or industry-level bargaining for those in the privileged fordist sectors of segmented labor markets. Even in the United States, however, the heyday of fordism witnessed attempts to complement military Keynesianism with a "Great Society" reinvigoration of the New Deal welfare state.

With the crisis of fordism on both sides of the Atlantic, however, we can discern a transition from systems of local govern*ment* organized around expanding, localized delivery of Keynesian welfare state functions toward a system of local govern*ance* organized around what, by analogy, can be termed a *Schumpeterian workfare role*. This role is quite novel. In economic terms, it attempts to promote flexibility, economies of scope, and permanent innovation in open economies by intervening more widely and deeply on the supply side of the economy and tries to strengthen as far as possible the structural competitiveness of the relevant economic spaces. In social terms, it subordinates social to economic policy with particular emphasis on labor market flexibility, structural competitiveness, and the impact of the social wage as an international cost of production (see also Jessop, 1993).

Developments in the global economy have radically altered the relevance of the typical fordist demarcations of economic space. National economies are no longer taken for granted as the main space or object of economic regulation, and the range of extra-economic conditions considered to be significant for securing economic competitiveness has been much extended. Thus, whereas the crisis of fordism initially led to attempts to reinvigorate the conditions for fordist accumulation at local and national levels, a consensus (valid or not) has since grown that the economic spaces most relevant to accumulation and the main extra-economic conditions for economic competitiveness have changed significantly. This is reflected in dominant economic discourses and the demarcation of spaces of accumulation. Alongside discourses of globalization, triadization, and so on (with their important supranational strategic implications), there is the alleged discovery (or, perhaps, rediscovery) of flexible industrial districts, innovative milieus, technopoles, entrepreneurial cities, learning regions, cross-border regions, global cities, and so forth (with their more localized strategic implications).

The impact of globalization, the growth of new core technologies, and the marked paradigm shift from fordism to postfordism[5] are also associated with a far broader account of the conditions making for economic competitiveness. It is now believed to depend on a wide range of extra-economic factors and thus to need a wide range of competitiveness-enhancing policies.

This dual reorientation has been reinforced insofar as regional and local economies have been seen to have their own specific problems that could be resolved neither through national macroeconomic policies nor through uniformly imposed meso- or microeconomic policies. This perception indicated the need for new measures to restructure capital in regard to these newly significant economic spaces and for new forms of urban governance to implement them. This is associated with demands for specifically tailored and targeted urban and regional policies to be implemented from below, with or without national or supranational sponsorship or facilitation. These tendencies are reflected at local level in a widening and deepening of initiatives in reskilling, technology transfer, local venture capital, innovation centers, science and high technology industrial parks, incubator units for small business, support for entrepreneurship, efforts to expand export markets, and so on. This in turn affects the definition of economic spaces. For, rightly or wrongly, they are seen as much more strongly socially or institutionally embedded and, perhaps consequently, as requiring more complex forms of regularization and governance than fordist forms of economic organization.

This is related to a shift in economic governance mechanisms away from the typical postwar bifurcation of market and state. Indeed, postwar forms of urban government that rested on this institutional distinction were often seen as ill-equipped to pursue new approaches and thus came to be seen as part of the problem of poor economic performance. Partly this is reflected (especially in Europe) in the search for supranational forms of govern*ment* to compensate for the deficiencies of local as well as national government; but it is also associated with the search for new forms of govern*ance* at all levels able to overcome the problems linked to pure market or hierarchical, bureaucratic solutions. Thus, we find new forms of network-based forms of policy coordination emerging— cross-cutting previous private-public boundaries and involving key economic players from local and regional as well as national and, increasingly, international economies. Subnational governments (including urban authorities) have been reorganized to promote economic regeneration in partnership with a range of local (or localized) economic, social, and political forces. This entails active, state-sponsored dispersion of local power from elected local authorities to a wide range of local (or localized) economic and political forces. This is intended to enhance the reorganized local state's strategic capacities in the ever more closely interconnected fields of social and economic policy making, and to help it cope with the far-reaching political repercussions of economic and social restructuring.

Although the increased significance of governance typically involves a loss of decisional and operational autonomy by state apparatuses (at whatever level), it can also enhance their capacity to project state power and achieve state objectives by mobilizing knowledge and power resources from influential nongovernmental partners or stakeholders. As regional and local states are becoming a partner, facilitator, and arbitrator in public-private consortia, and growth coalitions, however, they risk losing their overall coordinating role for and on behalf of local community interests and, thereby, part of their legitimacy. This problem is particularly acute where urban areas have active social movements with political agendas rooted in the continuing crisis of fordism or the economic and social pressures arising from more flexible, but also more insecure, postfordist economic order.

A final point to note regarding changing urban regimes is the enhanced role of the local state and local governance mechanisms in international economic activities. This is closely related to more general changes in the autonomy and capacity of national states due to the expansion of supranational intergovernmental regimes, local governance regimes, and transnationalized local policy networks. Duchacek was one of the first theorists to describe the expanding regional, provincial, and local government roles in international affairs. He referred to microdiplomacy (including overseas representation, promotion, lobbying, etc.) in such fields as economic interchange, environmental policy, and welfare (e.g., Duchacek, 1984). The result is a perforated sovereignty where nations are more open to transsovereign contacts by subnational governments and where international policy transfer between localities is likely to increase (see Cappellin, 1992; Church & Reid, 1995). This reinforces the importance of looking beyond increasingly artificial local boundaries in studying urban regimes and growth coalitions. It also poses the problem of increased vulnerability by allowing foreign actors to divide and rule different levels of policy making (see also Rycroft, 1990, pp. 218-219, 229).

Postfordist economies will be co-constituted through postfordist modes of regulation (see earlier), so there is no pregiven blueprint from which to derive appropriate forms of governance. Regulation theorists have long argued that new modes of regulation emerge as chance discoveries through trial-and-error search processes. This is reflected in continuing experiments to find new, more adequate forms of articulation of regulation and governance in response to narratives that ascribe part of the blame for failure and crisis on previous models of urban politics and local economies. It is hardly surprising that there is widespread experimentation with new forms of economic governance for new urban regimes and that there are always new fads and fashions for models (and their backers) that appear to promise success.

## Concluding Remarks

In the spirit of this volume, I have proposed an alternative research agenda intended to supplement and reorient rather than to wholly supplant the study of urban regimes. Thus, I have drawn on Gramscian state theory, the regulation approach, recent insights into governance, and some reflections on the spatial imaginary to present some key dimensions to urban regimes, their structural and strategic dimensions, their economic and ethico-political moments, and their embeddedness in a wider economic and extra-economic context. My primary concern with agenda setting and a firm editorial reminder not to stray beyond strict word limits have precluded any presentation of new empirical material (although my arguments have been shaped by ongoing research in Greater Manchester and the East Thames Gateway in England) (see, for example, Hay & Jessop, 1995). Given the principal aims of the present volume, however, this theoretical and methodological focus is probably justified. Accordingly I now want to note some key points for this alternative, strategic-relational approach to urban regimes.

Methodologically, I hope to have added to arguments presented elsewhere in this volume on the potential role of the regulation approach as a supplement to the still evolving urban regimes approach. But I have done so by putting a particular gloss on the regulation approach—one that notes its remarkable similarities to Gramsci's work on the state in its inclusive sense and, even more strikingly perhaps, his various reflections on the ethico-political and psycho-economic moments of economic regimes. It is certainly worth remarking the significant extent to which each paradigm adopts a strategic-relational approach and also places its specific theoretical object in its wider social context. Thus, whereas Gramsci examined the social embeddedness and social regularization of state power, the regulation approach examines the social embeddedness and social regulation of accumulation. For Gramsci, this meant examining the modalities of political power (hegemony, coercion, domination, leadership) that enable a historically specific hegemonic bloc (power bloc) to project power beyond the boundaries of the state and thereby secure the conditions for political class domination. Conversely, for regulationists, this involves studying the modalities of economic regulation (the wage relation, money and credit, forms of competition, international regimes, and the state) that regularize, discipline, and guide microeconomic behavior within limits that are compatible in given historical circumstances with the expanded reproduction of capitalism as a whole. In this sense both approaches are interested in the strategic selectivity of specific regimes (political or economic respectively) and their implications for class domination (likewise political or economic).

In bringing these approaches together, we can strengthen each of them and at the same time develop useful tools for studying the nature and succession of urban regimes. Gramsci's own work is itself marred by its gestural (if still theoretically tantalizing) treatment of the decisive economic nucleus of hegemony.

This neglect is often more serious in recent neo-Gramscian work. Regulation theory is one way to remedy this particular deficiency and has the virtue of intrinsic compatibility with Gramscian concepts. Moreover, as I tried to indicate above, it was in certain respects anticipated in Gramsci's writings. Conversely, the regulation approach is regularly criticized for its neglect of the distinctive dynamic of the state system and political regimes. Only a few theorists (mostly working outside the Parisian mainstream) have paid much attention to the state system. More generally there has been a one-sided concern with the various structural forms and institutions involved in the overall reproduction regulation of capitalism to the neglect of the many and varied governance mechanisms involved in the organization or self-organization of the complex web of inter-dependencies among these forms and institutions. Clearly this is a serious defect from the viewpoint of someone interested in urban regimes and suggests that in its predominant versions the regulation approach can at best *contextualize* the nature and succession of urban regimes rather than *explain* them (important exceptions regarding the state can be found in the work of German theorists working with regulationist concepts; e.g., Esser & Hirsch, 1989; Keil, 1993; Mayer, 1994). Last, the regulation approach has neglected the role of the ethico-political dimension to regulation and, in particular, the key role of economic discourses, the organic intellectuals involved in elaborating accumulation strategies and hegemonic projects, and their implications for the formation of economic subjects. The study of urban regimes needs to address such issues and link them to the current transformation of economic strategies.

Nonetheless, it is important to note here that neo-Gramscian theory and the regulation approach are still only complementary. They cannot be combined without further theoretical work to establish more detailed conceptual linkages and logical connections. In doing so, one must be careful to avoid any simple-minded reduction of urban politics to the needs of the economic base (let alone a base that is understood purely in economic terms). Moreover, as already suggested in my comments on governance theory here and elsewhere (see, for example, Jessop, 1995b), both the regulationist and neo-Gramscian approaches would benefit from more interest in issues of governance and metagovernance.

Two additional lessons must be stated. First, as already noted above, it would be wrong to attribute any (let alone all) of the changes identified in urban regimes theory simply to the effect of economic changes. The regulation approach is useful for contextualizing changes in the nature of urban regimes but cannot directly explain them. Due regard must be paid to how economic issues are first translated into political problems for action by the state *in its inclusive sense* and their solution is mediated by the structurally inscribed, strategically selective nature of political regimes. Moreover, as I have conceded earlier, by no means do all urban regimes prioritize economic development. Thus, many of the preceding arguments must be interpreted in terms of the need for alternative hegemonic projects to recognize that the viability of urban regimes depends in the last instance on the economy in the sense that adequate revenues must either be generated locally or redirected from more successful economic spaces elsewhere. Second, from a more state-centered viewpoint, it would be wrong to suggest that any of these trends is purely attributable to economic changes, however "integrally" or "inclusively" these are analyzed. For there could also be distinctive political reasons prompting state managers or other relevant political forces to engage in institutional redesign and strategic reorientation regarding local economic strategy (Jessop, 1992b, 1995a, 1995c).

## Notes

1. Anderson (1991) regards nations as "imagined" communities; states, regions, cities, and so on are likewise imagined entities.

2. Although Gramsci probably misattributes the idea of "determined market" to Ricardo, this is unimportant for present purposes.

3. American radical political economists who work on social structures of accumulation come even closer to the Gramscian concept of historic bloc. See, for example, Kotz, McDonough, and Reich, 1994.

4. On the mobilization of bias, see Schattschneider (1970). Similar ideas occur in the debate on the three faces of power between pluralists and elite theorists.

5. This reference to a paradigm shift does not imply that there has already been a transition from fordism to postfordism in the real economy. Apart from any conflation this would introduce between strategic paradigms and real economies, some commentators argue that a real transition has not yet occurred and that all that we can witness are various states of disorder.

# P
# A
# R
# T

# 2

# Reconstructing Urban
# Regime Abstractions

I n this section, the authors take a closer look at the urban regime
abstractions used to analyze the regulation of urban politics in a
global economy. Although in different language, Leo evaluates the
geographical variation in the roles of the central state and regulatory
processes and their interactions with the politics of urban planning and
economic development. Leo criticizes regime theory for failing to locate
urban politics in a national and global context. He argues that when the
central state takes a larger role in urban development processes (e.g.,
France, United Kingdom, and Canada), the power of large corporations
is somewhat neutralized or at least "the playing field is more level."
Stated differently, local governments left to their own (as often is the
case in the United States) are more vulnerable to the ever increasing
mobility of finance capital for large development projects and corporate
capital for offices and branch plants. At the same time, Leo finds that this
central state role also correlates with less citizen participation that he
speculates to be increasingly important. Thus, Leo believes that urban
regime theory must be reconstructed in a fashion that internalizes national and global political economics.

Whereas Leo's contribution to the reconstruction of urban regime concepts is to highlight differences in nation-states' role in urban redevelopment, Cox focuses on the reconceptualization and specification of mechanisms of cooperation in the governance of urban development and the spatiality of that governance. In the former, he focuses on the insights provided by considering transaction cost theory insights to explain cooperation within and between state and market actors. In the latter focus, he highlights the erroneous effects of viewing space as a mere backdrop or arena for political economic relations and social relations. Here he focuses on local dependence and its implications for cooperation and the transaction cost problem. Cox corrects the privileged role often given to local politicians and bureaucrats in coordinating the governance of local economic development strategies. He argues that such governance is more contingent than that recognized by urban regimes analysts.

Although Leo and Cox are mostly concerned with reconstructing urban regime theory, Painter considers the compounding burden of reconstructing an urban regime theory compatible with and useful for regulation theory. Painter suggests that a reconstructed urban regime theory can provide an approach that examines the mediation of regulation in and through specific social practices and forces, thereby resolving regulation theory's inability to nonfunctionally explain change. Compatible with my criticism of Stone's urban regime theory in the introduction to this volume, Painter argues that urban regime theory, as now constituted, explains regime formation and transition based on rational choice theory of behavior and thus is incompatible with regulation theoretic accounts. Painter proposes replacing the rational choice theoretic base of urban regime theory with Bourdieu's conceptualization of habitus, field, and strategy as a way of understanding the processes by which potential participants in a regime join and interact within a governing coalition. Here, as seen earlier in Feldman's chapter, an analysis of discursive practices becomes crucial in a reconstructed urban regime methodology. Finally, Painter's discussion of the habitus and fields of differing actors in a governing coalition and the subsequent stability or instability of the urban regimes closely resembles Jessop's turn toward Gramscian theory and the concept of hegemony discussed by Clavel (1995).

# 5

# City Politics in an Era of Globalization

## CHRISTOPHER LEO

L ocal politics has long been subject to influences originating in the global economy, but only lately have students of urban political economy begun to appropriate a language and a set of concepts intended to comprehend that reality. Much work remains to be done. In this chapter, I discuss urban politics in North America and Europe, with special attention to the politics of planning and development, ask how globalization is changing the political landscape, and inquire into the local political response to the challenge of globalization. This

AUTHOR'S NOTE: Thanks to Joe Painter, Mark Goodwin, and Bob Jessop for excep-
tionally helpful critiques of an earlier draft of this chapter, and to Mick Lauria for
imaginative, smart, tactful, and firm organization of our collective scholarly venture. I
also very much appreciate the research assistance that I received from Mike Gray,
Michelle Mathae, and Krista Boryskavich. And I am grateful to the Social Science
and Humanities Research Council of Canada, the University of Winnipeg, and the
University's Institute of Urban Studies for financial and other support. Although it
would be nice also to spread the blame for errors, I must unfortunately take all that
on myself.

question has not been given the attention it deserves. Much of the literature tends either to assume that globalization is bringing about a massive homogenization in the development of cities, or to observe significant political differences in different jurisdictions without reference to the literature that points to homogenization. In this review, I will show substantial homogenization and take a look at the dynamics that drive it, but I will also note local and national political differences that are capable of exerting significant influence on the way globalization affects city development. Comparing continental Europe with the United States, the greater involvement of the European national state in local politics produces a more nearly level playing field in bargaining between the state and capital. At the same time, I find a greater American penchant for public involvement in decision making about urban development, a penchant that is likely to prove an important asset in the local politics that is emerging under the aegis of globalization.

In looking at city politics, I am undertaking a study of urban regimes in the sense that I am concerned with the formal and informal "arrangements by which public bodies and private interests function together to be able to make and carry out governing decisions" (Stone, 1989, p. 6). I depart from most of the regime literature, however, in looking at city governance in a national perspective, that is, integrating politics at the regional (state and provincial respectively for the United States and Canada) and national levels of government into our conception of urban politics. Indeed, as noted, one main finding involves observations about contrasting roles of national governments in city politics.

My argument is that, in the politics of planning and development, cities are subjected to a wide range of pressures for homogenization, of urban built form, of urban physical structure, and of administrative and political arrangements, as a result of technological and economic developments that are global in scope. At the same time, urban governance asserts local particularity—local ideas of how cities should be governed, how they should look, and what the social milieu should be—in the face of these homogenizing pressures. I find that these two pressures—the push for homogenization and the assertion of particularity—are both important, both help to shape cities and the ways they are governed, but neither enjoys absolute dominance. However, although cities in Europe and North America face many of the same problems and often react to them in similar ways, important and deeply rooted differences remain and exert major influence. To undertake this investigation, and to bring globalization and the role of national governments in local affairs into the argument, however, we need to consider other literature that has dealt with these questions.

# Peterson: Primacy of Market Forces

The study of city politics in the United States has encompassed a concern with *globalization*—though it was not always designated by that term—since the early 1980s. Paul Peterson's (1981) contention that city politics is dominated by a common interest in pursuing policies that "maintain or enhance the economic position, social prestige, or political power of the city" (p. 20) was based on the observation that there is pressure on cities to tailor their policies to the requirements of those who might choose to invest in the local economy. His argument amounted, in effect, to a declaration that the global economy prede-termined a wide range of political outcomes. Although the ideological assump-tions and language in Peterson's book differed from that which dominate the literature about globalization, Peterson deserves credit for having been a pioneer in studying the impact of globalization on cities.

The dissent against Peterson focused on his argument that economic realities precluded extensive redistributive local politics, that the economic vulnerability of cities forced them to emphasize economic priorities over social ones, regard-less of political preferences. The response was that economic factors could not, by themselves, determine political outcomes. For example, Mollenkopf (1983) maintained that the political entrepreneurs who are the main source of urban policy are "driven by a political logic even more strongly than an economic one. It is in the interaction between the political and economic logics, and the conflicts between the two, that the guiding force behind the life cycle of pro-growth coalitions may be discovered" (p. 5). Likewise, Stone (1987) as-serted, "Local decision makers do not simply follow the imperatives that emanate from the national political economy. . . . Urban politics still matters" (p. 4).

The American reaction to Peterson's argument did not highlight an aspect that seemed questionable from a Canadian perspective, and even more so from a European one: his assumption that there is a clear distinction between national and local policy spheres and correspondingly clear differences in fiscal capacity. These differences between the conditions of national and local governance were a cornerstone of Peterson's argument. He explicitly stated that nation-states exercised substantial control over the flows of capital and labor over their boundaries and that they could therefore pursue social or redistributive objec-tives far more readily than local governments, but that local states, unable to regulate these flows, were forced to emphasize policies designed to attract investment over those designed to ease social distress (Peterson, 1981, pp. 17-30).

The idea that a clear distinction exists between national and local spheres—assuming it makes sense in the United States—certainly breaks down when one tries to apply it to Canada. A few examples will suffice to establish a point that is pursued in more detail elsewhere (Leo, 1995b). In cities across Canada, government programs have had a major impact on inner cities. These programs have ranged from construction of downtown malls and public facilities, through heritage preservation and the construction and renovation of housing for all income levels, to numerous job creation and training programs and government funding for public and semiprivate social agencies. In Peterson's terms, these programs constitute a jumble of both developmental and redistributive programs. If we took Peterson's theory seriously, we would have to predict that the former were generally local-government initiatives whereas the latter remained primarily a domain of either provincial or national governments.

In fact, in case after case, these initiatives have involved either joint ventures of two or more levels of government, participation by provincial and national governments in local initiatives, or local participation in federal or provincial initiatives. And local government participation did not, à la Peterson, particularly lean toward developmental as opposed to redistributive ventures. The city governments of both Toronto and Vancouver, for example, have played a major role in the development of subsidized housing, and in Winnipeg the city government was an equal partner in a trilevel program, the Core Area Initiative, that contained substantial redistributive elements. By the same token, both Toronto's Harbourfront development and Vancouver's Granville Island were federal government initiatives that were primarily developmentally oriented. Indeed, Harbourfront was widely criticized for an undue emphasis on its developmental elements, at the expense of redistributive programs (Leo, 1995b; Leo & Fenton, 1990), thereby placing the federal government squarely in the center of the type of "local" political debate that occupies a prominent place in Peterson's discussion.

The question of whether Peterson's insistence on fundamental differences between the capacities of local and national governments truly holds in the United States is perhaps better left to American commentators. What is obvious to an observer from outside the United States is that Peterson's theory is tied to a specific set of political conventions and fiscal capacities. Assume a major change in local taxation powers or a national government willing and able to play an activist role in local affairs and the theory loses much of its force.[1] These different conditions—conditions tending to invalidate Peterson's theory—exist, in varying degrees, in both Canada and Europe. In Canada, a wide-ranging literature finds Canadian political culture to be more oriented than American to state intervention in society and the economy.[2] One impact on urban politics of these differences is that during the 1970s and 1980s, while the U.S. federal government was reducing its involvement in local affairs, the Canadian government continued a wide range of interventions, including the establishment of state-owned corporations to carry out local initiatives (Leo, 1995b; Leo & Fenton, 1990).

In Europe, especially on the continent, the differences identifiable in a comparison with the United States are considerably more striking even than they are in Canada. As we will see in more detail later, an emerging comparative literature observes levels of national intervention in local affairs that are all but unthinkable in either the United States or Canada (Body-Gendrot, 1987; Klaassen & Cheshire, 1993; Molotch & Vicari, 1988; Savitch, 1988; Vicari & Molotch, 1990). Savitch (1988), characterizing the French system of urban development, says, "Government is the dominant force in initiating and supervising major development. . . . That presence is embodied in the actions of the national government . . . local governments and public corporations" (p. 136). Body-Gendrot (1987) states flatly, "France . . . [has] a strong-state tradition . . . State power is not fragmented by . . . provinces, counties, cities, and neighborhoods" (pp. 126-127).

In short, Peterson was right to insist on the importance of the global economy, but his characterization of that influence has not traveled well because the theory is applicable only to a specific set of political circumstances, circumstances that more nearly reflect American rather than Canadian or European city politics. It is grounded, not in fundamental features of any nation's political economy but in temporally and spatially limited distinctions between different levels of government (Cox, 1991a, 1991b). It has remained for regulationists to forge a theory that roots the relationship between economics and politics in more fundamental realities.

# Regulation Theory

Regulation theory seeks its understanding of the fundamental nature of urban social, economic, and political life in an investigation of how technology, the economy and social relations, including formal social organization of all kinds, evolve in relation to each other. Using this method, it is possible—to take a relatively clear and simple example—to consider how the transition from an industrial system oriented to mass production of standardized consumer products to one more oriented to small-batch production of more nearly customized products is related to the erosion of the welfare state. Similarly, relationships can be identified among the erosion of the welfare state, the removal of many barriers to trade and the rapid development of technology that allows instant communication and ever easier, worldwide movement of people, goods, information, and ideas (Harvey, 1985; Scott, 1988; Storper & Walker, 1989). It is important to emphasize that these are mutual interrelationships: Economic change helps to produce alterations in technology and changes in social organization, social transformation is involved in the production of economic and technological change, and so forth (Jessop, 1990a, 1995; Painter & Goodwin, 1995).

Thus—again sticking to relatively simple examples—regulationists have drawn on Henry Ford's insight that well-paid workers would be able to afford the cars they produced and argued that an industrial system oriented to mass production produces an influential source of political support for a welfare state that provides income and other social supports, thus ensuring a high level of mass consumption, hence the ability of ordinary people to buy cars and a wide range of other mass-produced consumer goods. It follows that a relative decline in the importance of mass production of standardized products removes an important part of the political basis for the welfare state and plays a role in forcing government withdrawal from or reduction of many former activities.

The state's ability to play its accustomed role is also undermined by the development of communications technology, which allows both the relatively easy relocation of industrial operations from one country to another and the movement of capital in the blink of an eye. A political implication of this change is that the state loses much of the political power that flowed from its capacity effectively to control industry and finance within its borders. That loss of control implies reduced ability to pursue traditional taxation and regulatory policies. Here we catch an echo of Peterson (1981, pp. 25-29) who, writing in the late 1970s, proceeded from the premise that the national government enjoyed the power that comes from controlling the movement of capital and labor over its borders, but that local government did not. Regulation theory argues that his premise was wrong or only temporarily right. Today it is obvious to any observer of politics that not only cities but also nations are repeatedly subjected to the whims of the global economy. Just as developers, operating on a world scale, demand concessions for the construction of essentially identical shopping centers in cities around the world, so industrial and financial organizations demand similar tax concessions from Tory, Liberal, and Social Democratic regimes alike.

Thanks to its grounding in more fundamental realities, regulation theory shows us that Peterson's insight about political weakness in the face of economic

necessity has a far wider application than he suspected. It also provides escape
from the realm of parochial theorizing, based on temporally and spatially limited
circumstances, and offers some theoretical grounding for the study of comparative urban political economy on an international level. Thanks to influences
originating in regulation theory, it is now becoming clear to students of urban
political economy that local politics needs to be understood, not only in a
national political and economic context but also in the context of the global
economy and global society. Now let us consider how a recognition of the
importance of the global economic, social, and political context can affect our
understanding of the politics of urban planning.

## Global Homogenization

It is nowhere more evident than in the politics of urban planning that local
and national political practices and cultures interact with development norms
that have become global. Or, to put it another way, norms and practices based
in the global economy have a homogenizing effect on local and national political
cultures. Among the homogenizing forces are the spread of technologies that
have an impact on urban form, for example the private automobile, which
confers greater mobility on a large number of people and encourages lower-
density development; computer and communications technology, which allows
for further decentralization of residence and workplace; and the subway, which
provides a transportation infrastructure that encourages the development of
high-density commercial and residential concentrations around subway stations
(Blumenfeld, 1967). The result is similar suburban subdivisions, "edge cities,"
and high-density developments in widely scattered cities.[3]

A more personalized homogenizing influence is the demands of developers
who seek to make their developments conform to international architectural
norms and technical standards. Often, meeting those demands necessitates
compromises with, or abandonment of, locally set standards of design or
administrative and political practices. Another homogenizing stimulus is the
example of other jurisdictions. Administrators and politicians are constantly on
the alert for ideas originating elsewhere that can be applied locally. Two
examples of ideas that have spread and had an important impact on cities in
different parts of the world are urban development corporations as an administrative mechanism for downtown redevelopment projects and design controls as
a way of creating a supposedly unique character for each neighborhood—a

"uniqueness" that is often replicated in Chinatowns, coffee house districts, and downtown malls in city after city. Yet another homogenizing influence can be seen in international agreements and organizational structures, such as the North American Free Trade Agreement and the European Union, which compel many changes in local as well as national administrative norms.

Much of the literature on globalization, especially the earlier literature, has tended to stress homogenization. For example, in discussing the impact of the global economy on cities, Logan and Molotch (1987) classify all American cities into only five types, according to their economic function: headquarters cities, innovation centers, retirement centers, and so forth. Noyelle and Stanback (1984), approaching the same problematic from a different ideological perspective, identify 4 categories, albeit 11 subtypes. Storper and Walker (1989) consider how cities are shaped by industrial growth, and Scott (1990) inquires into how cities are shaped by the division of labor. They all, of course, allow for differences among cities but all concentrate on differences created by variations in the economic forces acting on them. The cities become, in effect, urban units whose differences are seen primarily in conditions and changes that are global in scope. Fainstein (1994) looks at the influence of property development on London and New York City and strongly stresses similarities in the property developments she investigates. Sorkin (1992) offers perhaps the ultimate expression of this emphasis. Complaining about the impact on cities of technology and current development styles, he said, "The new city . . . eradicates genuine particularity in favor of a continuous urban field, a conceptual grid of boundless reach" (p. xii).

Studies that emphasize the way global economic and cultural influences exert pressure for homogenization have contributed a great deal to our understanding of cities today, and the politics of planning offers much evidence in support of their perspective. Canadians, for example, have watched since the 1950s as their cities changed to resemble the way American cities had already begun to look before World War II. The first American-style suburb was Don Mills in Toronto, which was being built in the early 1950s (Sewell, 1993). The Place Ville Marie office tower complex in Montreal, an enterprise of the New York developer William Zeckendorf, was completed in 1962 (Collier, 1974). In 1965, the Toronto Dominion Centre, an office tower complex designed by the eminent Chicago-based modernist architect Mies van der Rohe, was being built in Toronto (Lorimer, 1978). Today both suburbs and downtown cores across the country resemble those of American cities, the overall urban structure more compact (Goldberg & Mercer, 1986, chap. 7) but the style indistinguishable at ground level, as witness the fact that many American movies are shot on location in Toronto, Vancouver, Montreal, and other Canadian cities. More recently, the influences have spread to Europe. La Défence in Paris is a forest of high-rises and London's Covent Garden, a long-established wholesale food market and working-class neighborhood, has become a fashionable mall surrounded by upscale shopping, restaurants, and entertainment. The area has more historical depth than any city in North America can manage, but otherwise it is patterned on many similar developments across the Atlantic. La Défence and Covent Garden are only two of many similar examples.

All that is widely understood, but, although America is clearly the dominant cultural influence, the influences do not all flow in one direction. We can see that, once again, by looking at recent North American urban development from a Canadian perspective. A marked difference between Canadian and American suburban development was that, although American suburbs were largely the product of a decentralized building and development industry, the Canadian government deliberately fostered the establishment of large developers, who operated in a more highly regulated environment than did their American counterparts (Bacher, 1993; Carver, 1948; Cullingworth, 1987, chap. 12; Leo, 1995b). For some time, the Canadian adoption of American patterns of urban design continued under the auspices of divergent organizational principles, with the American development industry continuing to be relatively decentralized, and the Canadian one highly centralized. In the 1980s, however, as downtown revitalization came into vogue in the United States, so did the need for developers who could take on large projects and cope with regulatory intricacies. Canadian developers, accustomed to operating on a large scale in a highly regulated environment found a market niche ideally suited for them. Already in the early 1980s, Cadillac Fairview was undertaking a $1 billion development in a five-block area of downtown Los Angeles, Daon was developing several office towers in various U.S. cities and had become the second-largest publicly owned real estate firm in North America, and Olympia and York's development of Battery Park City at the foot of Manhattan Island was under way (Newman, 1982, pp. 29, 110, 112, 222). To be sure, the story of Canadian developers in the United States has not always been good news for the developers. Both Daon and Olympia and York collapsed in the turbulent land markets of the 1980s and early 1990s. Nevertheless, Canadian developers have become a factor in American real estate markets.

One reason why a centralized, regulated development industry found a niche in the freewheeling American market is that the American market is not as freewheeling as it once was. In part, this is due to the very nature of downtown redevelopment, which often necessitates large-scale development, a challenge to which the giant Canadian developers were well suited. In addition, the vicissitudes of redevelopment in a previously densely developed area include extensive interaction with municipal authorities, a challenge for which Canadian developers' experience in a more regulated environment prepared them. But it is not just a matter of large projects requiring extensive consultation over curb cuts, utility corridors, water connections, zoning regulations, and a host of other matters, significant as that is. The public-private partnerships that were a recurring theme of American downtown revitalization required local governments to display more flexibility and adopt a more entrepreneurial spirit than had been customary in the past (Frieden & Sagalyn, 1989). But they also demanded of the developer a willingness and ability to engage in extensive and complex dealings with government officials. Canadian developers were well placed to succeed in the American market because that market was becoming more regulated/reglemented (see Goodwin & Painter's distinction in this volume). In other words, in this respect, the United States was becoming a bit more like Canada, or Europe for that matter.

The move to a more regulated/reglemented environment extends far beyond downtown redevelopment and, indeed, is so far-reaching that I can only touch on it here. A growing interest is developing in a variety of state interventions designed to combat the ills of both urban sprawl and inner city decay. In considering these regulatory/reglementory initiatives, it becomes more and more difficult, with the passage of time, to draw systematic distinctions between Canadian and American approaches. A sizable American literature on growth controls demonstrates that there has been a growing interest since the 1970s in local or state government control of new development, despite mixed results.[4] Design controls and programs of bonuses and linkage are being instituted in many cities on both sides of the border and are spreading from city to city (Altschuler, Gomez-Ibanez, & Howitt, 1993; Dreier & Ehrlich, 1991; Goetz, 1991; Herrero, 1991). It is no longer as clear as it once was who is ahead in this move toward more political intervention in the development process. For example, Vancouver has had very stringent design controls since the early 1970s, undoubtedly more stringent than any others in Canada, and probably in the United States as well. Portland, Oregon, and the state of Oregon began also in the 1970s to establish not only an extensive regimen of design controls, but also strict regulations governing development at the urban fringe (Abbott, 1983; Abbott, Howe, & Adler, 1994; Knaap & Nelson, 1992). Oregon's Urban Growth Boundaries are more stringent and comprehensive than any currently in force in Canada.

A conspicuous remaining difference between the two countries is that U. S. regulation measures, as well as the opposition to them, are more oriented to court action than those in Canada. A dispute that, in the United States, might be resolved in the courts would, in Canada, more likely be fought out on the floor of city council, before a provincial municipal board, or around the provincial cabinet table. Undoubtedly other differences could be identified, and such differences are significant, but my main point is that many American local and state governments have become more interventionist in recent decades in their regulation of urban development and that many of the same methods are being used in the two countries to respond to problems of urban sprawl and downtown decay. Clearly, from these examples, significant convergence is taking place, and, at least in some respects, the similar problems cities face are evoking similar responses in a world where technological, economic, and even political influences are global in scope.

# The Politics of Particularity

If much of the globalization literature stresses, or even assumes, homogenization of cities in the face of global pressures, an emerging literature also cultivates an eye for local particularity while placing local politics in a global context.[5] If we cull from that literature, and add to it some material from a growing stock of cross-nationally comparative studies of urban politics,[6] we can gain some insights into the degree to which politics "still matters" (Stone, 1987, p. 4), despite the overwhelming importance of global economic forces.

As I have throughout this chapter, I will focus on the politics of urban planning and development. Given the practical necessity of tailoring my investigation to the available literature in a field that is still young, I will pay special attention to the politics of inner-city redevelopment.

It is useful to think of the bargaining over urban development as involving three parties: the state, capital, and citizens.[7] I will begin by comparing the relationship between the state and citizens on the two continents and then look at the more crucial relationship between the state and capital. The database is slender, and my findings are preliminary. My intention is to contribute to an overview of the differences between European and North American city politics. Others have already made contributions to that attempt (Savitch, 1988; Savitch & Kantor, 1994; Fainstein, 1994), but no single, widely accepted approach has emerged, and the time for its emergence is probably still well in the future. What I found in my investigation is nicely summarized in a comment of Goodwin, Duncan, and Halford on the impact of the global economy on local governance:

"Market criteria have to be imposed on wide areas of state activity, and this process is one of conflict and struggle, not automatic technical change" (1993, p. 85).

## The Role of Citizens

Case studies comparing Paris with New York show a marked difference in the degree of legitimacy accorded to direct participation of citizens in the process of urban development. For example, Savitch (1988), in comparing New York with Paris, classifies New York City's decision-making process as "a blend of corporatism and pluralism" (pp. 59, 60). He finds corporatism in both cities, but it is New York's distinctive feature that "non-established groups, with few resources and no privileged access" can wield pivotal influence, especially if the corporatist decision-making process, involving "top politicians . . . business, labor, interest groups, and community boards" gets bogged down. His strongest example of the potential of citizen opposition is the case of Westway, a massive Manhattan road-building scheme, which was first proposed in 1969, encountered fierce resistance, and had still not been built at the writing of his study. In Paris, by contrast, where "mobilizing corporatism" holds sway, "mass opposition to the political elite is surprisingly limited. . . . Relatively few decisions evoke the ire of the citizenry" (p. 137). In three case studies, Savitch found citizen opposition once, and then only "at the eleventh hour." In that case, the opposition, as he demonstrates in some detail, did not significantly influence the outcome.

Savitch's findings are supported in Body-Gendrot (1987), a study of a rare Parisian case of relatively successful neighborhood resistance to a major development that came only after a large part of the neighborhood had already been redeveloped with "35-floor towers." The author prefaces her account with a characterization of France as a "strong-state society" where "particular wills" are "not as well-protected as in the United States" (pp. 126, 129-130). It is clear, from both her account and Savitch's three case studies, that the position of ordinary citizens in Parisian development politics is very different from that in New York. Two other case studies (Fainstein, 1994) confirm New York's side of that dichotomy while making it clear that, although the legitimacy of citizen involvement is well-established, its ultimate effect is often marginal. Broadly, her findings coincide with the others.[8]

In these studies New York and Paris are poles in the North American-European dichotomy, and clearly they respectively reflect *modi operandi* that are observable elsewhere on the two continents. At the same time, there is intracontinental variation—perhaps, given the slender database, more than anyone suspects. Some of that variation is visible in case studies of British and Canadian urban development politics, and it suggests that those nations occupy a middle position between the poles represented by New York and Paris. Both Savitch (1988) and Fainstein (1994) include London in their comparative case studies, and their findings are interesting.

Savitch, having found a corporatist-pluralist hybrid in New York and mobilizing corporatism in Paris, characterizes London politics with the term *liberal corporatism.* In brief, his argument is that in London, as in New York but not in Paris, citizen input is a regular feature of development politics, but the response of the state is different. In New York, he observed, citizen groups remain independent of the state throughout the decision-making process, battering the state from outside when they disagree with proposals. In London, representatives of protesting groups that prove their mettle in the political arena are incorporated into the decision-making process, thereby giving them a direct role in decision making while neutralizing them as opponents.[9]

For example, in the chic commercial redevelopment of the rooted working class neighborhood of Covent Garden, the Covent Garden Community Association (CGCA), "a fiery and effective pressure group" with a radical political style "embarrassed public officials with its skill and thoroughness" (Savitch, 1988, pp. 219-222). The state's response was to form a representative body, elected from the community. As a result, "What had once been an opposition was suddenly converted into a collaborative organization with special access to decision makers." The upshot was a redevelopment that was humanized in scale and appearance and more protective of the neighborhood's environment and history than the original proposals, and that incorporated some benefits to local residents. At the same time, a working-class neighborhood was gentrified and transformed into "one of the most chic locales in Central London." The example suggests a much greater seriousness about citizen input in London than in Paris but a very different style than in New York.

In four comparative case studies of development projects in New York and London—two in each city—Fainstein (1994) produced broadly similar findings.[10] In all four cases, citizen participation occurred and was accepted by the authorities, as a matter of course. And although Fainstein does not report any actual co-optation of community representatives onto government bodies, as Savitch does, it is notable that in both the London cases the citizen groups were given financial assistance by the government, whereas no government finance for citizens was reported in either New York case. On this evidence, as on Savitch's, if Paris and New York are the poles, London is somewhere in the middle of the spectrum.

Comparable observations could be made about the Canadian situation. Citizen participation in urban development decisions is a legitimized, though sporadic, part of the political process (Sancton, 1991, p. 473). In the 1970s, there were a number of high-profile citizen rebellions against development initiatives. Among the best known of these cases was the Marlborough Avenue campaign against Marathon Realty's Summerhill Square development in Toronto, in which local residents fought a tenacious battle before city council and in the court of local public opinion to win a modest list of concessions from the developer (Granatstein, 1971). Also in the 1970s, a series of citizen uprisings against urban expressway schemes proved partly responsible for the fact that most Canadian inner-city neighborhoods remain largely unscathed by multilane highways. A major factor in the greater success of such initiatives in Canada than the United States, however, was the paucity of federal government funding for expressways (Leo, 1977).[11] In the 1980s, the Downtown Eastside Residents' Association, representing the residents of Vancouver's Skid Road area, won several remarkable concessions from local government in the area of cooperative housing and tenants' rights (Ley, 1994). These and similar efforts, however, have been sporadic and have been preceded and followed by periods of dormancy.

Perhaps the most striking example of institutionalized citizen participation in Canada is the development permit approval process in Vancouver, in which notice of any new development is posted in the neighborhood affected, details of the plans are made available to the public, and then a process of consultations and hearings takes place in which the developer is required to respond to the concerns of citizens and to conform to a rigorous set of design guidelines (Leo, 1994). This, however, has the earmarks of the co-optation evident in the Savitch (1988) data on Covent Garden. On the available evidence, Canada, like Britain, appears to be a mid-Atlantic case: Canadian urban politics is clearly more amenable than French to citizen initiatives, but seemingly less so than that of the United States, or at least of New York City.

## The State and Capital

However important citizen participation may be in the political scheme of things, it is also clear, from a variety of evidence, that the actual impact of citizen initiatives on the shape of urban development is limited. At its most potent, citizen participation can score some successes, usually partial ones. At its least influential, it is window dressing or is altogether absent. The same cannot be said of the influence of capital. Therefore, I turn now to the more interesting and more crucial question of how the relationship between the state and capital differ in European and North American politics of downtown redevelopment.

The available literature points to a striking contrast that directly affects the interactions, and the respective roles, of capital and the state in the process of urban development: The European national state has a markedly more prominent role in local decision making than the American one does. For example, in an Italian case study, Vicari and Molotch (1990) argued that growth machines do not exist in Italy as they do in the United States because of national regulation of land use, a highly centralized party machinery, "systematically integrated with lower echelons," and the fact that "local government is not financially or legislatively independent of the central state" (p. 619). All of these factors, according to Vicari and Molotch, combine to insulate local government from the kind of development pressures that are a familiar feature of North American politics. The influence of capital is wielded at the national level, especially through party machinery, where land development interests must vie with industrial, financial, and other political interests for attention (Molotch & Vicari, 1988). Several Parisian case studies (Body-Gendrot, 1987; Savitch, 1988) paint a comparable picture of French politics. There, a strong, activist, centralized government and a powerful national bureaucracy play a more directive role in development than any found in North America: "It is not uncommon for French developers to behave as supplicants to a powerful class of technocrats" (Savitch, 1988, p. 134). In France, as in Italy, pressures from land developers are only one of a wider array influences on a state marked by a high degree of integration between national and local levels.

Savitch's (1988) studies of New York predictably find decision making on development matters lodged primarily at the local level and far more exposed to direct pressures from developers. He noted that decision making is marked by "the city's perceived need to attract investment, increase the value of land, augment tax revenues, provide jobs" (p. 59). Fainstein (1994), drawing conclusions from a series of case studies comparing New York and London, added that both Mayors Koch and Dinkins "felt compelled to respond to every notice by a major firm that it was considering a move to New Jersey with a counteroffer" (p. 164), and observed, "In both London and New York . . . projects had government sponsors, but the willingness of government to offer direct subsidy to the developer was far greater in the latter city" (p. 166).

Savitch's British cases offer an interesting comparison and contrast. Although the central government's creation and subsequent abolition of the Greater London Council, both by fiat, have earned it a formidable reputation for high-handedness, Savitch's studies of development politics found a less central-ized situation in Britain than in France. The national government is actively involved in development decisions, but Savitch (1988) finds a relationship more marked by central-local competition there than in France, and his case studies picture a robustly competitive political arena, with local governments that do not appear as agents of the center. He commented, "If it wishes, the central government can override the local opposition with impunity, but it rarely does . . . relative to interest groups, French government is far more centralized and monolithic than British government" (p. 203, n. 4). From that evidence, it would appear that here, as in the matter of citizen participation, Britain occupies a mid-Atlantic position.

Canada occupies a different, but similarly ambivalent, mid-Atlantic position. Canadian cities, like American ones, are directly exposed to developer pressures because they are primarily responsible for decisions relating to land use. City governments negotiate with developers, as Toronto's local authorities did in the Marlborough Avenue case described earlier, and if their bargaining position is weak, they may find themselves making massive financial and material concessions, as Edmonton did in the Bank of Montreal and Eaton Center developments (Leo, 1995a) and Winnipeg did in its dealings with Trizec (Walker, 1979). Although the federal government does not exercise supervision over local land use, it has, as I have noted already, involved itself very significantly in local development, through a wide range of measures, including federally owned corporations that sponsored such redevelopments as Toronto's Harbourfront and Vancouver's Granville Island; participation in a trilevel bureaucracy that managed the Core Area Initiative, a multifaceted program for the revitalization of Winnipeg's inner city; and federal programs to finance the relocation of downtown railway properties and the construction of convention centers (Leo & Fenton, 1990). Canada's nondirective but interventionist federal government places Canada, like Britain, in a mid-Atlantic position in this comparison.

## Conclusions

In this chapter, I have considered what happens to the study of the politics of urban development and planning when it is placed in a global context. I contrasted Peterson's narrow view of globalization's impact on urban politics, which is grounded in temporally and spatially limited distinctions between different levels of government, with the broader conception flowing from regulation theory, which directs our attention to the need to view local politics, not only in a national political and economic context but also in the context of the global economy and global society. I then examined the politics of urban development and planning and considered, first how pressures that are global in scope are having a homogenizing effect and then how politics asserts itself to forge important local differences in the responses to globalization. It was not, however, local politics per se that made the crucial difference; rather it was the role of the national state in city politics. Drawing on case studies, I found that the strong national state presence in urban politics evident in France and Italy was associated, on one hand, with a political environment that tended to be unreceptive to grassroots participation in urban development decisions and, on the other hand, with more state control and less clout for developers in urban development decisions. In the United States, by contrast, I found the opposite situation: a more receptive environment for grassroots participation and developers who were better placed to exert direct influence on urban development. In Britain and Canada, a more complex, mixed picture, a mid-Atlantic state of affairs, was discernible in the data.

Another way of summarizing my findings about national power and the character of local politics is to say that, in the United States, power is more fragmented than it is in Europe, and that this fragmentation is not just a matter of a different way of doing things, but also has consequences for political outcomes. That the national state has less say over local affairs enhances the power of one group, the development industry, in relation to other social groups and this, in turn, affects the way cities are governed and ultimately the way they look and the kind of society that develops there. These findings confirm the argument in Newton's (1975) classic article, in which he argued that fragmentation of power affects social class, political structure, and the distribution of public goods.

My findings are based on careful, detailed case studies, but it must be stressed that it is an emerging, and therefore slender, database. Any conclusions we can draw from it are necessarily tentative and hypothetical. Such tentative findings can play a useful role in mapping out directions for research, however, even if some of them are subsequently invalidated. It is appropriate, therefore, to end this investigation on a somewhat speculative note, offering clues to future research by drawing out some of the possible implications of current findings.

## National and Local States

My findings suggest that Peterson's delineation of clearly distinguishable national and local policy spheres and fiscal capacities does not reflect the reality of European city politics. That lack of salience is visible in the fact that discussion of Peterson's works has been primarily a North American, and especially American, preoccupation. It is also visible in differences between European and American cities and city politics.

In the New York cases, I found comparative local autonomy in land development matters, with local authorities that had a great deal of control over their own land use policies, whereas the national government tended to keep hands off. This is a common state of affairs in the United States. Decisions about land use are largely a result of bargaining between local authorities, the developers or other corporate interests that are proposing new developments, and any citizens that become involved in the decision-making process. Two sets of power relationships emerge from this constellation of political forces. One power relationship is that between corporations, some of them with deep pockets and far-flung international interests, and local governments that may be very vulnerable. Their vulnerability can be heightened by the ever-increasing mobility of finance and of corporate offices and industrial branches.[12] In many cases, corporations deliberately ignite bidding wars between municipalities to see where they can get the most generous assistance and the most liberal land use regimes.

The other power relationship typical of decisions about urban development is that between the local state and citizens, sometimes representing the neighborhoods most affected by a development proposal, and sometimes constituting broader coalitions. The case studies suggest that direct citizen participation in the making of particular development decisions is a more important factor in the United States than in continental Europe. All these political conventions are reinforced by characteristic features of American political culture, including a widely held belief in the importance of allowing private enterprise to operate with a minimum of hindrance; a belief in local autonomy, which takes the concrete form of the widely observed principle of local "home rule"; a suspicion of "big government," especially if that government is a national government; and a suspicion of powerful, interventionist government, whether local, regional, or national.

The continental European examples that were available for my study show, by contrast, much more involvement by national-level politicians and officials in the process of making decisions about land use. Greater national government involvement in urban affairs tends to shift the determination of urban policy from competition between corporations and relatively weak municipal governments to a bargaining process involving national governments with corporations that can wield influence at that level, many or most of which are not in the land development business. That interurban competition for development is not as prominent a feature of European as of American urban politics stems partly from the frequently active role of national governments in urban development, a role that necessarily includes centralized allocation of development opportunities, together with at least some presumption that each city will get a share.[13]

Likewise, that decay of inner cities is a less pressing problem in Europe than in the United States is undoubtedly traceable partly to the stronger national political role in cities, which, again, is bound to imply, in some degree, a national commitment to the health of those cities.

The more powerful role of the national state not only enhances the state's position in its bargaining with corporate interests but also creates a situation that, from an American perspective, might appear as undue subordination to the state. From a European perspective, a more apt formulation might be that the involvement of national-level politicians in land use decisions produces something closer to a level playing field in the bargaining between the state and capital than is found in the United States. At the same time, direct citizen participation is much less of a factor in continental European urban politics than in the United States.

## Implications for Regime Theory

My findings also point to the need for a reevaluation of regime theory in light of the insights that come from looking at urban politics in a national and global context. Although regime studies do not make Peterson's mistake of arguing as if the division of functions between levels of government characteristic of the United States represent fundamental attributes of all government, neither do they offer much help in dealing with the reality of complex intergovernmental patterns of social production. Indeed, regime theory has paid scant attention to the global context of local politics. Although there have been exceptions, in practice, regime research has focused on urban case studies, usually those of a single city (Shefter, 1985; Stone, 1989; Whelan, 1987; Whelan, Young, & Lauria, 1994), sometimes comparing two cities (DiGaetano & Klemanski, 1993a; Lauria, 1994b), and generally not viewed in a global context.[14]

Placing local politics in a global and national context introduces considerations that have not, in any systematic way, been integrated into regime theory. If local policies are being produced, now by one or another level of government, now by complex interactions among two or three levels, what happens to regime theory? How do we conceptualize the political base for such decision making? Will we have more than one regime per city, one locally based, another with a regional political base, and perhaps a third relying on some combination of local, national, and regional support? Perhaps, instead of regimes, it will prove more useful, in the first instance, to think of coalitions, overlapping to be sure, but each constituted to address a different set of policy concerns: one focusing on economic development, another on housing, a third on environmental problems, some entirely local in their composition, others more broadly based. At the end of the day, however, there is a panoply of decision makers who, among them, determine the course of events in any given city. Some means ought to be available to conceptualize this state of affairs, but this means would necessarily take us beyond regime theory in its present state.

This is not to suggest that regime theory has outlived its usefulness. It has produced formidable insights and offers important and otherwise unavailable lessons about what is involved in building a political coalition and holding it together, and about how the task of coalition-building changes depending on whether its purpose is downtown development, neighborhood improvement, or frugal use of taxpayers' money. These lessons greatly clarify the conditions for successful political action, whatever level of government is brought into the engagement. But they do not capture the complexity of the urban governments we find when we view them in a global context, or undertake cross-national comparisons. Deciding how our understanding of urban regimes fits into the context of national politics and the global economy is an important question for future reflection and research.

### Applications

The comparative study of urban political economy and of the various responses to the pressures of globalization is not only important to academics. It has immediate practical significance as well. The ubiquity of the globalization pressures on cities means that they have more and more to learn from each other's experiences. A globally oriented analysis of city politics can play a useful role in that process of learning and adaptation. The fate of American inner cities may be an object lesson to America and the world, but there are also more edifying lessons to be learned from the way U.S. cities are governed. If urban politics in a wired world is to have any prospects of constructively channeling the forces of globalization, it is unlikely that this will be accomplished, in the long run, without involving the public far more actively in decision making than they were in some of the European examples we have considered in this chapter. Usable models are much more likely to be found in the United States where a tradition of individualism and populist politics has forced urban politicians and administrators to develop habits and techniques of public involvement that could be applicable and useful elsewhere.

It was possible for Peterson, and has also been so for much of the regime literature, to make important contributions to our understanding of urban politics while proceeding on the assumption that the national state was a discrete entity, cleanly separable from local politics. In a global perspective that is not possible because once we start comparing how the global influences of cities helped shape those cities, we cannot carry out meaningful comparisons without considering the full range of things governments do, or refrain from doing, in cities. Our unit of analysis must become not just the local regime but all of the forces that help to shape political decision making in and for the city, whether they originate at the local, regional, or national level. What is more, the influence of such supranational governmental and quasi-governmental institutions as the European Union and the North American Free Trade Agreement are sure to be felt in the administration of cities, and they too will have to be included, at least peripherally, in our understanding of urban politics. It is a fascinating challenge, one that poses interesting and complex analytical problems.

# Notes

1. Peterson tacitly acknowledges this in his final chapter when he proposes extensive federal intervention in local affairs as a way of overcoming the city limits he points to in the rest of the book.

2. See Frye (1982), Goldberg and Mercer (1986), Horowitz (1978), and Lipset (1990). Garber and Imbroscio (1992), in arguing that Canadian city politics offers the necessary conditions for the operation of growth coalitions, seek to turn the debate in a different direction, but do not dispute the findings cited here.

3. Regulation theory reminds us that the global economy's homogenization of local planning systems—together with local political resistance to these influences, which I will discuss in the next section of this chapter—are only one part of a far more complex set of relationships. The global economy not only homogenizes, it also differentiates, through uneven development of different cities and different neighborhoods, for example, and through cycles of urban decline and regeneration. Likewise, politics not only resists homogenization, it also homogenizes. The policies associated with the names of Ronald Reagan and Margaret Thatcher played an important role in changing the character of economic policy making around the world. The example of Reaganism and Thatcherism also illustrates that the economy, in addition to acting on politics, is also acted on by politics. A recognition of all these complexities, and many more, does not prevent us from identifying relationships, however, and drawing out their implications, one at a time, as I am doing in this chapter.

4. Among the useful sources are Baldassare (1986), Downs (1988), and Gottdiener (1983). Altshuler et al. (1993) offers a good overview of measures in the United States designed to make some of the proceeds of new development available for other purposes.

5. Much of this literature grows out of sociological and geographic studies. Students of politics have shown less interest (Goodwin et al., 1993; Lauria, 1994b; Leo, 1994, 1995a; Pickvance & Preteceille, 1991; Sassen, 1991; Stoker, 1990).

6. A joint British-American conference at the University of Bristol in 1994 added a good deal of impetus to a field that had been developing slowly in previous years (Chen & Orum, 1994; DiGaetano & Klemanski, 1994; Fainstein, 1994; Ferrer & Van Til, 1994; Hambleton, 1994; Jauhiainen, 1994; Jezierski, 1994; Lauria, 1994b; Molotch & Vicari, 1988; Pickvance & Preteceille, 1991; Savitch, 1988; Savitch & Kantor, 1994; Stoker, 1990; Vicari & Molotch, 1990).

7. The term *citizen* is being used here in the general sense of "resident" or "the public," a common usage in the North American political science literature. The connotation that citizens should be viewed as holders of a particular status or as persons with particular rights and obligations (as in Isin, 1992) is not intended here.

8. Fainstein could be interpreted as arguing the ultimate marginality of citizen participation somewhat more strongly than Savitch.

9. It is interesting to note that in the rare case of successful citizen opposition in Paris reported by Body-Gendrot (1987), leaders of the neighborhood organization—which benefited from the advice of some residents who had inside knowledge of politics and administration—made a point of declaring and maintaining their independence of political parties, precisely so that they would not be co-opted.

10. Fainstein's book actually contains six case studies in all, but because neither of the others—Docklands and Battery Park—involved new development in the middle of established neighborhoods, they are less helpful in comparing the role of citizens in the making of development decisions.

11. As well, a demonstration effect was operating. The opportunity to watch as numerous American cities were carved up by expressways undoubtedly affected the attitudes of Canadians. In any event, it is not being argued here that Canadian politics is more amenable than American to citizen involvement. On the contrary.

12. By the same token, local governments' vulnerability can be minimized by a city's superior attractiveness as a location for business (cf. Kantor, 1987, p. 496; Leo, 1994).

13. The national government's doling out of its share of development to each city and region is also a conspicuous feature of Canadian politics (Bakvis, 1991).

14. Depending on exactly how regime theory is defined, the following could be cited as exceptions: Horan (1990), Fainstein (1990), Feagin (1985), and Smith and Feagin (1987).

# 6

# Governance, Urban Regime Analysis, and the Politics of Local Economic Development

KEVIN R. COX

## Context

A central objective of urban regime theory is to provide some conceptual leverage in understanding the politics of local economic development. One way this happens is through regime theory's emphasis on the importance of cooperation. This is seen primarily as a problem between public and private sectors: The division of labor between state and market and the subsequent need for cooperation across that divide is insisted on. As Stone (1993) has recently written, "Urban regime theory assumes that the effectiveness of local government depends greatly on the cooperation of nongovernmental actors and on the combination of state capacity with nongovernmental resources" (p. 6). But beyond this affirmation of interest and reference to the public/private divide, there is not much to go on. For a start one would like to know more about the

AUTHOR'S NOTE: I would like to acknowledge the very helpful comments that I received on an earlier draft from the editor (Mickey Lauria), Marshall Feldman, Bob Jessop, and Joe Painter.

mechanisms of cooperation. Just what, for example, are the various ways in which cooperation across the public-private divide can be secured? There is a literature on this, and it can usefully be brought to bear on the problem.

A second area of neglect concerns the role of space. This arises in several related ways. Arguably, one reason the urban regime literature has not problematized the variety of mechanisms of governance that actually exists is that cooperation as an issue has been assumed rather than understood. A major reason that cooperation is an issue is that a number of agents with stakes in local economic development, private and public, are place dependent in some of their social relations. This limits the range of other agents with whom they can interact and creates a variety of problems that the urban economists have grasped as those of spatial externalities and monopolies: precisely the sorts of conditions that, in more abstract terms, can explain the functional significance of governance mechanisms. These place-dependent social relations, moreover, are in their spatial reach quite variable and can in no way be reduced to the jurisdiction of a particular local government. Yet there are marked tendencies in the urban regime literature toward privileging those jurisdictions as the essential arenas for bridging the public-private divide.

Finally there is the question of the public-private divide itself. If indeed the metropolitan is a meaningful arena for organizing for local economic development, why shouldn't local governments enter into arrangements, understandings perhaps, with each other as well as with more private agents? Moreover, why does local government have to be involved at all or, to put it somewhat less radically, as involved as it is? Granted that there are problems of cooperation in promoting local economic development, to what extent are they problems that necessarily call for cooperation with local government rather than some set of relationships between private agents? The boundary between polity and economy is always shifting. There seems no a priori reason for assuming that the boundary must be in a position that privileges local government as not merely involved but as the central nexus of organizational efforts.

This chapter is divided into two major sections. I start out by discussing the issue of cooperation for local economic development in relatively abstract terms and the three issues identified earlier are addressed in turn. In the second major section, these arguments are illustrated with two case studies. The first looks at the problem of organizing water and sewerage provision in the Columbus metropolitan area. Water and sewerage availability are key conditions for attracting in new investment. In the Columbus area provision is on a metropolitan scale but not through a metropolitan agency, posing interesting issues for the relationships between local governments.

The second case study highlights the role of relations of trust in mediating inward investment into metropolitan areas. Local governments are involved in this, but contrary to common impressions, their role is quite subordinate. And once again this is a process of cooperation that occurs on a metropolitan, to some degree, suprametropolitan, scale. This refers to organizing for local economic development in the four largest metropolitan areas of Ohio, including that of Columbus.

# Cooperation and the Politics of Local Economic Development

## Urban Regime Theory and Governance

The idea of governance is certainly no stranger to those active in urban regime analysis, though its use has yet to acquire buzzword status. Yet there is a taken-for-grantedness in the way it is mobilized and in the innocence of the broader literature on the topic that is disarming. This is not to say that the use of the term *governance* is inappropriate. References to the formation of coalitions and to relations of trust in the cementing of structures of cooperation across the public-private boundary do reassure. The concern with governance in the context of local economic development is also germane. Given the present way in which it is constituted and the partial way in which it approaches the problem, however, urban regime theory can provide only very limited purchase on governance. A useful collection on the problem of governance is titled *Markets, Hierarchies, and Networks* (Thompson et al., 1991). Of the three mechanisms so identified in that title, urban regime theory seriously draws only on the third. A more complete view has to include the other two mechanisms and in their relation with one another.

Part of the reason for this partiality is a failure to move beyond descriptive references to various forms of public-private collaboration to a reasoned questioning of why social forms of that nature actually emerge. Just what are their more abstract preconditions? Why, in the case of urban regime theory, is it reasonable to expect relations of trust between, say, different developers and the local planning department? And why might in other circumstances markets suffice?

Much of the literature on governance starts from the failure of markets to provide the necessary social coordination among agents: "necessary," that is, for their particular objectives. But we should not ignore the role that markets or quasi-market relations actually perform in modes of governance. Under certain ideal conditions markets are self-regulating; under less than ideal conditions they require supplementation by other mechanisms—mechanisms like "hierarchy" or the various network forms. Moreover, due recognition should be given to the role that those mechanisms that can most accurately be described as "quasi-market" play, particularly in the public sector as well as in its relations with private agents.

Exchange relations in the public sector are common enough, and in the relations of various state agencies with firms, labor unions, and the like. The literature on "horse trading" and "pork barrel legislation" and Anthony Downs's analysis of party competition provide obvious documentation of this. The exchange of favors between neighboring local governments is an almost every-day feature of local politics. The political economy literature, which tries to understand government "outputs" in terms of the classical demand and supply relations and competition, is also a large one. Likewise in the competition for inward investment there is an exchange of various incentive packages offered by local government for, in effect, an expected future stream of public tax revenues and employment. But in contrast with what is more typically subject to market analysis, in the cases discussed here the exchange relation is one in which the magnitudes of the respective payoffs are uncertain. Cost-benefit analysis, including various estimates of multiplier effects, can be used by local government to justify offering incentives to an incoming firm but these estimates are subject to notoriously wide margins of error.

But the fact of some choice for the parties to the exchange in these cases lends additional credence to the validity of treating them as at least quasi-market in form. Firms do trade off one locality against another, just as some localities can exercise a degree of discretion about the sorts of firms they want to attract: clean, white-collar versus smokestack industries, and so forth. The imposition of land use controls on new developments by local governments testifies to similar forms of quasi-market tradeoffs.

A central point in the governance literature, however, has been the recognition that market relations do not always bring about the exchanges desired by agents. Where competition is imperfect, where, that is, those party to exchanges have few alternative agents with whom they might transact, then there are serious possibilities of malfeasance and all manner of opportunistic and exploitative behavior. Reasons for the nonsubstitutability of transacting agents vary. Supplying a particular firm with some service or component can require investment in sunk costs that could not be applied to production for any other client if the purchasing firm should demand a renegotiation of the price or withdraw some of its orders. Where a firm or organization, like a government agency, transacts repetitively with some other, the parties involved can gain information that could be used to enhance respective leverages, perhaps in an asymmetric way. In some instances the lack of alternatives is rule defined: local governments have only one county with whom they can deal.

The economist Oliver Williamson (1975) was one of the first to approach this problem, though he was much more concerned with the firm than with other organizations like agencies of the state. Where markets cannot provide an effective counter through the provision of readily substitutable agents, the solution would be what Williamson terms hierarchy or administrative fiat: internalizing the transactions within the firm and subjecting the agents involved to a set of binding rules. Transactions that are uncertain in outcome, recur frequently, and require substantial transaction-specific investments—money, time, or energy that cannot be easily transferred to interaction with others on different matters—are therefore, according to Williamson, more likely to occur within hierarchically organized firms whereas the more straightforward, non-repetitive exchanges that require no transaction-specific investment will be between firms and consummated in the market. In other words, and according to Williamson, whether transactions are internalized within the firm or external-ized in the form of arms-length relations with other firms depends on the relative efficiency of the two solutions. Relative degrees of vertical integration or disintegration in firm structure hinge on this consideration.

The contributions of Granovetter, the sociologist, have also been important. Granovetter (1985) focused on relations of trust as an alternative means of achieving social coordination, again in a context where more decentralized exchange relations of a market or quasi-market nature would not suffice. Granovetter's arguments are framed to a very substantial degree with respect to Williamson's arguments and, indeed, those of the new institutional economics in general. The critique is, essentially, that Williamson's transaction cost theory is undersocialized. In its basic assumptions, transaction cost theory is, like the tradition of neoclassical economics of which it is a legatee, methodologically individualist. Though Granovetter does not exploit this essential insight to the point that he might, clearly, Williamson's base point of a world of individuals who come together only in exchange is a chimera, and that exchange itself depends on the presence of other forms of social structure: not just the relations of trust that Granovetter emphasizes but also the rules that protect private property rights and that are laid down and enforced by the state.

But, and perhaps because he does not press his point to the degree that he might, Granovetter also sees complementarities between his own contribution and that of Williamson. For where markets were, for whatever reason, inadequate for the task of social coordination, administrative fiat or hierarchy and networks of trust relations could be functional substitutes for each other:

> We should expect pressures toward vertical integration in a market where transacting firms lack a network of personal relations that connects them or where such a network eventuates in conflict, disorder, opportunism, or malfeasance. On the other hand, *where a stable network of relations mediates complex transactions and generates standards of behavior between firms, such pressures should be absent* [italics added]. (Granovetter, 1985, p. 503)

It is useful to pause at this point and note how these arguments apply to the problem of cooperating for local economic development. To some, albeit limited, degree the idea of trust has been featured in the urban regime literature. Given the localized character of land development projects and the way in which there are often at any one time competing projects at different locations within the local government's jurisdiction it is evident that there have to be understandings between the developers and the local government. For a variety of reasons projects are scheduled over time, though not always in a very planned manner. Unless developers have some degree of trust in local government that those overlooked in the first round of investments will have first priority later, there is a danger that the matter will become highly politicized. One consequence can be an escalation of public costs as the mobilization of allies results in the need for side payments (for an example, see Hoxworth & Thomas, 1993).

But the broader literature on governance also appears useful. The idea of hierarchy is, arguably, underexploited in the urban regime literature. As a mode of handling transaction cost problems, the public sector analogue to Williamson's vertical integration is the amalgamation of formerly separate state agencies, or the creation of a new agency to take over functions formerly handled by separate agencies. Proposals for metropolitan integration have as a goal this form of internalization of externalities. New agencies at a metropolitan scale—airport commissions, mass transit authorities, or even revivified counties to which municipalities have ceded some of their powers—can accomplish the same objective and are common enough as vehicles for the mediation of local economic development.

There is also a close relation between Williamson's formulation of the problem of cooperation and theories of the state. The elimination of the transaction cost problems inhibiting market formation is commonly seen as a, if not the, purpose of state authority. Even granting the weaknesses of the transaction cost approach this is interesting because it allows the urban regime literature to be situated with respect not just to ideas about governance but also to the burgeoning literature on regulation. In this regard it points to a discourse, which though undeveloped from the standpoint of the problem of organizing for local economic development is complementary and therefore thoroughly apropos.

Like regulation theory, the focus of regime theory is the economic. But beyond that, the relation is very weak. Urban regime theory seems much more concerned with the bottom-up construction of networks of relations between local government and private sector actors than with the top down imposition of norms or the provision of conditions for some normative convergence at the microlevel.[1] This does not mean that an approach to the politics of local economic development could not be more regulationist in character. Local governments do resolve problems of uncertainty, fulfill roles in the division of labor that would otherwise go unfulfilled, and this promotes local economic development. The whole apparatus of land use control is a case in point: externalities are internalized as a result of a state assumption of authority. This allows transactions to occur, say between landowners and developers, that would not otherwise have occurred.[2]

This suggests that a more coherent understanding of organizing for local economic development would require attention to the various complementarities between top-down and bottom-up forces. This should proceed not just with relations within more local spheres of government. What is equally needed is more attention to the relations between various agents, including local government itself, with stakes in local economic development and more central branches of the state. There is, for example, a regulatory division of labor between those more central branches and the more local, but this is mediated as much by bottom-up as by top-down forces.

## Problems of Spatial Context

Mechanisms of social coordination have their necessary conditions. In the previous section I suggested that one reason for the failure of urban regime theory to situate itself more centrally with respect to that body of work resided precisely in its seeming aversion to interrogating the necessary conditions for those mechanisms. Urban regime theory has shown itself sensitive to mechanisms of cooperation but not to the issue of why those mechanisms come into being, are reproduced, transformed, and contested, and so forth.

One reason for this is the problems that urban regime theory has had in incorporating ideas about space and how people relate to space. Urban regime theory makes reference to space and it delimits a space as the sphere of certain types of political activity, that is, "the urban." But space remains something of a backdrop to the analysis rather than a set of relations that is actively mobilized by agents, which perhaps constrains them, and which they actively construct. Space is a passive, mere arena rather than something that is used, that constrains in various ways, and that is formed.

I argue here that the spatial relation is a condition for mechanisms of cooperation; as far as local economic development is concerned, the spatial relation is a necessary condition for the sorts of market failures, exploitative situations, and transaction cost problems isolated by the governance literature. Without certain spatial relations cooperation would not be a problem—there would be no reason to organize for economic development at all.

But if we are talking about spaces of cooperation, spaces within which agents find the relations—trust, hierarchy, quasi-market exchange—that facilitate their coordination or within which they feel the need to try to transform, or even construct anew, those relations, then the scope of that spatial arena is also called into question. A perusal of the urban regime literature suggests that, at the most generous reading, this issue that has not been considered a problem. For the most part, the arena for the formation of urban regimes is defined by the jurisdiction of a particular local government. Yet there is no a priori reason why the relations through which cooperation for local economic development is secured should stop at municipal boundaries. In metropolitan areas, the metropolitan provides a significant arena of action not just for developers, utilities, and other private agents with interests in local economic development but also for the local governments themselves.

One starting point for addressing these questions comprises the spatial relations of those agents who have major interests in local economic development.[3] Elsewhere Andrew Mair and I (1988) have defined these relations as those of "local dependence." If one considers those organizations, or firms, that typically come together in American cities to form growth coalitions, and that have major interests in local economic development, one property they have in common are various place-specific social relations that immobilize them, that limit their spatial alternatives to particular local, metropolitan, or possibly regional economies. To the extent that local governments depend for their revenues on a local tax base, property, sales, income, or whatever, rather than on intergovernmental grants, then their revenue prospects depend on the growth of a specific local economy. It is not surprising, therefore, that in the United States, local governments have been such ardent advocates of local economic development policies.

Local government is not alone in having these stakes. Other agents, like developers, the utilities, the banks, have also, historically at least, been important and by virtue of similar limitations in their sociospatial relations. Developers tend to be locked in to particular markets by virtue of a local knowledge and reputation that are not portable elsewhere and that take a long time to acquire, so that moving into another market imposes substantial opportunity costs. Only recently has the lifting of historic limits on intercounty and interstate branching allowed banks some measure of independence from local economic outcomes. And utilities have been affected by a combination of circumstances: not just a limitation to relatively small multicounty but also substate service areas and long-lived large capital investments. As a result of confinement to a localized market, this means that much rides on making projections of future demand materialize.

As a result, the problem of organizing for local economic development is rife with nonsubstitutabilities of the sort identified in the literature on the transaction costs problem. Banks have had to deal with particular local governments while the utilities continue to face that dilemma. Local governments have to deal with contiguous local governments. Developers face a similarly attenuated array of possibilities for finance, water and sewer lines, and the like. Simply put, if conditions in a particular metropolitan area are adverse, if exchange partners are exploitative, offer poor service, and so forth, there is no way, given the local dependence of these agents, that they can move elsewhere in search of the market discipline that would provide more favorable conditions.

As with the earlier discussion of Williamson's and Granovetter's approaches, this does not imply any bottom-up construction of modes of cooperation. I do not believe that a methodologically individualist, rational choice approach to the problem is at all defensible. This type of construction can certainly seem to take place, and there are suggestive examples (for a good example of this, see Lorenz, 1993). But existing modes of social coordination invariably form an important condition for new ones. Granovetter's personal relations can provide a matrix for developing trust relations in spheres to which they did not originally apply. The role of family as a mode of organizing business relations is exemplary. More apropos to the present case, a history of understandings between local governments may be the necessary precondition for their willingness to subordinate themselves to some new, more centralizing rule-setting body at a metropolitan level.

Where place-dependent social relations end and structures of cooperation begin is not always clear. To some degree it may be the relations of trust, the particular understandings that have been developed that actually lock agents into particular local arenas: not just the reputations of developers built up slowly over time but also those of the civil servants and even elected officials in neighboring local governments. Particular locality-specific rules regulating exchange can provide agents with such competitive advantages that, again, there are no feasible locational alternatives elsewhere. This means that we should be wary of any simple reduction of mechanisms of governance in particular arenas to relations of place dependence. Those mechanisms of governance can, in fact, be an important condition for place dependence. In other words, the problem needs to be approached in a much more dynamic way akin to the sorts of formulations of the structure-agent problem that have recently become common: Giddens's (1984) structuration theory and Bhaskar's (1979) transformational model provide instances.

The problem has important scale features. There are social relations that embed agents in particular places. Sometimes these are structures of cooperation in the way I have just outlined. Sometimes, however, they are clearly quite separate, and structures of governance emerge through a variety of often mutually presupposing processes: ones of intentional construction, ones of an evolutionary selecting out of existing structures for new purposes, and so on. But whatever the causal nexus these are social relations—more accurately sociospatial relations—that cannot be reduced to the scale of the municipality—he scale, that is, that has been the focus of urban regime analysis.

The relations necessary to particular agents can be specific to particular neighborhood contexts as in the case of developers trying to organize an appropriate environment of adjacent land uses. The relations can also be at scales larger than local government—embracing in some instances the whole metropolitan area. The market for industrial sites and for shopping malls is metropolitanwide so municipalities are drawn into a struggle with each other for tax base. Movements within the metropolitan area ignore jurisdictional boundaries. As a result, there is always the possibility of fiscal exploitation of one local government by the residents of others. The most obvious of these is the so-called suburban exploitation of the central city. This type of problem, moreover, can deter large investments, which will have spillover effects for other, noncontributory local governments: investments like new modes of mass transit, cultural facilities, or convention centers.

Alongside these sorts of relations, externalities, and spatial monopoly effects are also, as an empirically verifiable matter, structures of governance at similar spatial scales. I intend to illustrate this at greater length in the case studies but we can note in passing such well-known instances of metropolitan hierarchy as the New York Port Authority and the South Coast Air Quality Management District. As will be shown later, relations of trust also span interjurisdictional boundaries. And at more neighborhood levels, local government often delegates its powers to public-private corporations: urban renewal authorities, central city improvement districts, and the like. But the important point is that, strictly as a matter of our more abstract understandings of space economies, there is no reason why local government jurisdictions should provide the arena for the mobilization, construction, and transformation of structures of governance.

## The Public-Private Divide

Something of an article of faith in the urban regime literature is the centrality of the division of labor between the state and capital: If there is to be local economic development, the roles of state and capital need integration. This integration is to be achieved through building coalitions and creating understandings by local government. Again, from the standpoint of organizing for local economic development, I think that this is a very restrictive statement of the problem.

We have already seen that local governments can indeed feel the need to organize with other local governments. Privileging the activities of local government in general also seems to be problematic: local government seems to be the primum mobile around which the structures of governance in urban areas revolve. But I would suggest that there is no a priori reason why that should be. Other local stakeholders apart from local agencies of the state are equally, perhaps in some cases more, concerned with local economic development and the forms of governance that mediate the establishment of conditions favorable to it.

Governments can facilitate conditions for the rent appropriating activities of developers, but developers can take their own steps to ensure the right sort of environment for themselves. Privately funded freeway interchanges are not figments of the imagination and to the extent that the resources are available, the strategy can obviate any need to cooperate with other developers in sharing public spending. More common is the use of hierarchy—as where developers seek to maximize their rents by assembling large tracts of land. Within these tracts the various externalities they create can be internalized. This may be because they develop the tract themselves or because they make rules for those developers purchasing land there. There is, in other words, no need to rely on the land use control mechanisms of local government.

In yet other instances, local government may well be involved but the initiative has been taken by private sector agents. Developers can take the lead in attempting to solve local congestion problems within metropolitan areas. To some degree they can make progress simply by working with other private sector agents—cooperation among major local employers to encourage car and van pools, for instance. Physical infrastructural change is usually different. It can proceed so far through business donations toward the construction of a railroad overpass, for instance. But ameliorative projects, such as widening a freeway or inserting a new interchange, usually need the involvement of local government and as many local governments as are affected by the freeway.

In other words, governance in urban areas, organizing for local economic development is much more open both with respect to means and with respect to ends than is envisaged by urban regime theory. It is "open," for example, in terms of the arena within which the crucial social relations occur; it is "open" in terms of the repertoire of structures of governance and the various hybrid forms subsequently arrived at; and it is "open" with respect to who takes the initiative—not just local government but often private firms, nonprofit organizations, for example. I now want to illustrate these arguments with two case studies.

# Case Studies

## Organizing for Infrastructural Provision

The City of Columbus has highly developmentalist policies. The growth of the city and of its tax base have been central objectives for at least the last 50 years. The local Chamber of Commerce, currently known as the Greater Columbus Area Chamber of Commerce, has found it easy to share these goals. Creating a physical infrastructure to facilitate that growth has accordingly been a priority.

A major part of that infrastructure consists of water and sewerage systems. Given the suburbanizing tendencies of so much postwar growth, the timely extension of water and sewer lines on the urban periphery has been an important priority for developers. But this area of provision has not been a source of conflict, latent or otherwise, within the city itself. Creating a coalition of forces among developers, banks, utilities, and the various city departments involved behind the city's water and sewer policy[4] has not been difficult. Rather, the principal problems of cooperation have been external, with those local governments enjoying some contiguity with the City of Columbus. In order for us to understand why exactly that is, the problem needs to be framed with respect to the broader context of the jurisdictional fragmentation of the American city.

Tax base is the major developmental stake for local governments. Largely for this reason, they promote and support initiatives aimed at bringing in new basic sector investment. But from the standpoint of the firm making the investment, what may be attractive may be less one particular local government in a metropolitan area and more the metropolitan area as a whole. So although local economic development initiatives may be supported, even financed, there is no assurance that they will result in payoffs for the local tax base rather than for the tax base of some other local government in the metropolitan area. As a result of suburbanizing tendencies in business location, this is obviously a problem that has been exacerbated for central cities. This concern to capture the tax base effects of new investment has keened the interest of central city local governments in mitigating strategies like jurisdictional integration with suburbs and in annexation. In this context one can understand the City of Columbus water and sewerage policy.[5]

Water and sewerage have been used as tools to facilitate annexation to the City of Columbus at the expense of annexation by suburban local governments. The policy had two aspects, introduced sequentially. First, in 1954 the provision of water and sewerage by the city was made contingent on annexation. Until then the county had negotiated a number of sewer and water line extensions into unincorporated areas that were intended to cater to outlying residential and industrial developments. But after 1954, this was no longer possible. In exchange for providing these services, the City of Columbus demanded annexation.[6]

After 1964, a new and important complement to this policy consisted of water and sewer agreements with independent suburbs. In these agreements, Columbus agreed to provide the service in question, either water or sewerage or both, at agreed rates and to serve the suburb and a limited area into which the suburb might be able to expand by annexation in the future. These agreements were used strategically by the City of Columbus to limit annexation possibilities for the independent suburbs and to clear the way for its own expansion between and around their jurisdictions. The advantages that the suburbs gained in exchange for yielding this advantage to the city were lower water and sewerage rates, which were possible because of the city's ability to exploit economies of scale.

In these policies, the city has been highly successful. The possibility of cheap water and sewerage in exchange for a substantial forfeit of expansion possibilities has been attractive to the independent suburbs. By 1989, 90% of the county's population was served by the Columbus Division of Public Utilities. From a 7% share of the total land area of the county in 1940 the city had by 1989 increased its share to 34%. The city has also managed to surround several independent suburbs, testifying to its ability to stunt their growth to its own advantage. Furthermore, by virtue of its contiguity with unincorporated areas, the city retains the ability to expand still more in the future.

Historically, the independent suburbs were put in a situation where service contracts with the City of Columbus were attractive. They wanted to expand but lacked the infrastructural capacity to do so. In the meantime, Columbus was annexing unincorporated land that these suburbs might well have annexed themselves if they had had the necessary capacity. In addition, the State Board of Health was concerned about the impact not just of expanded sewage disposal facilities but also of the existing ones. This was because a number of the suburbs disposed sewage into rivers, which then flowed through Columbus, and Columbus obtained some of its water supply from one of them.

The major source of opposition to the annexation policy turned out to be the suburban school districts. Until 1955, municipal and schools annexation moved together. When unincorporated land was transferred to the city, that same area was transferred to the Columbus City School District. This had two effects in the suburban school districts. One, it reduced their tax bases. Two, it reduced student enrollments to the point at which the viability of the systems might be questioned.

This opposition was extinguished by a move in 1955 to decouple city and schools annexation. No longer would schools annexation be mandatory on city annexation. A State Board of Education was established with the power to determine whether or not there should be a transfer of land from one school district to another subsequent to city annexation. Several rules were established that, in effect, protected the suburban school districts from territorial attrition. In particular, there could be no transfer if the supplicant school district had a lower per capita property valuation than the school district from which land would otherwise be transferred. This meant that suburban school districts no longer opposed the city's annexation policy. Even when independent suburbs entered into service contracts with the City of Columbus that restricted their own annexation, the likelihood was that the school district's territorial integrity would be preserved subsequent to annexation by Columbus. But it also meant that annexation by Columbus School District lagged far behind that by the city. Indeed, by 1980, a remarkable 40% of the City of Columbus was in suburban school districts.

More recently, however, the balance of forces supporting these rules governing water and sewerage provision has been upset. Since the late 1970s, the city's ability to impose its terms on local governments and developers in areas yet to be annexed has suffered something of an eclipse. The major reason for this has been a change in the rules governing the relation between schools and city annexation—the rules that were so important in removing the roadblock of suburban school district opposition back in the 1950s.

The immediate context for this change was the introduction of busing for racial balance in Columbus City Schools in the late 1970s. This affected residential location decisions. The Columbus School District lost favor to the advantage of the suburban school districts. Pupil enrollments in suburban school districts expanded though not all of this expansion was white. The net effect has been an increasing black pupil composition in the Columbus schools: by 1980, almost 50% of the students. The school board also encountered increasing difficulty in passing operating levies, though to what extent this was a result of a declining tax base with the decline of residential property values in the school district or simply a result of a decline in the proportion of households with children in the public schools is unclear.

Meanwhile, much of the expansion outside the Columbus School District took the form of a housing boom, which, although within suburban school districts, was also in the City of Columbus. This placed the annexation policy of the City of Columbus in a new light. The Columbus School Board recognized that there was a connection between the decoupling in 1955 of schools and city annexation and their own problems of finance and falling enrollments. If these areas had been annexed to the school district when they had been annexed to the City of Columbus, not only would there have been fewer places for disgruntled parents to move to, but the racial composition of student enrollment and the fiscal position of the school district would have been very different at the time busing for racial balance was introduced. The school board moved to reopen the question of schools annexation.

In the suburban school districts with large areas in, or large numbers of students coming from, areas in the City of Columbus, the result was intense opposition. A major concern was retroactive annexation: the possibility that those City of Columbus areas that had not been annexed to the Columbus School District at the time of their city annexation would now be transferred. This was a concern for parents in those areas as well as for the suburban school boards concerned about reduced pupil enrollments and tax revenues.

Interestingly, the housing developers were also bitterly opposed to retroactive annexation. Busing for racial balance had meant that the Columbus School District was no longer attractive to the developers of new housing.[7] The boom was largely in areas that were in the City of Columbus and also in suburban school districts. Housing was being built and land purchased, in anticipation that the territorial affiliations of these areas would remain undisturbed.

Retroactive annexation did not take place. According to the agreement reached in 1986 among the different school districts in Franklin County, no land that had already been annexed by the City of Columbus but not to Columbus Schools would be transferred. In the future, however, city annexation would be automatically accompanied by schools annexation. The decoupling of the two forms of annexation that had been in place since 1955 was now undone.

Although from the standpoint of the developers and the school districts, this was preferable to retroactive annexation, it still was not very palatable. Sooner or later developable land in the suburban school districts, serviceable through contracts with the City of Columbus or actually in the City of Columbus, would be used up. It seemed likely, therefore, that in the not-too-distant future developers would be faced with the problem of building housing in Columbus School District and trying to market it to people who weren't very interested in locating there.

For school districts the danger was different but no less threatening. Many school districts included large areas of unincorporated land. In the past, annexation by the City of Columbus had posed no threat to student enrollments and to their property tax base. But in the future, it would. Families in those unincorporated areas were likewise alarmed. As a result of the annexation policies of the City of Columbus and limitations on the expansion areas of the suburbs contracting with the city for water and sewerage, these families too could look forward to a future in Columbus City School District: not very attractive either for parents or for those with large tracts of land that they hoped eventually to sell for development.

The result has been to open up the annexation policy of the City of Columbus to a renewed scrutiny. Developers have sought alternatives to annexation by the City of Columbus. Suburban school districts have embraced the search for these alternatives and been joined by households in those areas of the county that remain unincorporated. Consequently the city has been pressured to alter the balance of advantage in water and sewer contracts negotiations—particularly, the negotiation of the expansion areas for the contracting suburbs.

The forms of pressure have been twofold. First the specter of sources of water and sewerage service alternative to the City of Columbus has been raised. Unincorporated areas lying to the north of the county are, hypothetically at least, serviceable by a rural water and sewage utility in the next county, Del-Co in Delaware County. In addition, and on the eastern side of the county, one township, in an effort to stave off attrition through annexation, has established its own water and sewage treatment facility, despite significant opposition from the City of Columbus. This has been looked to both as a source of service for developers in the eastern part of the county and as a model for developers elsewhere.

The second form of pressure has been the city-township merger. Many independent suburbs with contracts with the City of Columbus abut unincorporated land, areas that currently fall under the jurisdiction of the townships. Contracts with the City of Columbus limit the possibilities of suburban annexation of township land and hence, from the standpoint of township residents and landowners, protection from annexation by the City of Columbus and, therefore, by the city school district. The merger of a suburb with a contiguous township, however, avoids this limitation.

The outcomes of these pressures must be considered against the backdrop of the two distinct aspects of the annexation policy. Land that is unincorporated can be serviced by Columbus if it is annexed. Alternatively, it can also be serviced by Columbus if it is annexed by a suburb with a contract with the city. Whether it can be annexed by a suburb, however, depends on whether or not it falls within the suburb's expansion area as agreed to in the service contract.

Developers close to cities lacking a service contract have therefore pushed for provision through such a contract with an expansion area sufficiently large as to include the land they want to develop. Where a suburb already has a contract and is threatening merger with an adjacent township the result has been a renegotiated contract: one that allows a larger expansion area but has a stipulation that a merger will void the contract. This means that there has been a shift in emphasis in the Columbus annexation policy so that annexation into the city is less likely, particularly for residential developments. Cooperation between city and suburbs has been reached once more, therefore, but with a different balance of advantage.

## Mediating Inward Investment

The case just reviewed exemplifies the role that hierarchy can play in achieving cooperation for local economic development on a metropolitan scale. Hierarchy there has been achieved not by integrating suburbs within the central city or by creating metropolitan water and sewer districts to replace provision by the municipalities but, rather, by contractual relationships. Instead of something akin to the vertical integration of the business world, therefore, the result has been analogous to the relation between a contractor and certain captive buyers.

I now want to illustrate my arguments about the significance of mechanisms of cooperation on a metropolitan scale with a second case that shows the significance not so much of hierarchy as of trust relationships. This case also illustrates the way in which initiative can lie less in the hands of local government and more in those of private agents. The concrete concern here is with the mediation of inward investment. Cities invest in infrastructure and try to improve the local business climate by offering various tax incentives. But the actual negotiations with firms considering a location in the area is a quite separate process—one in which local governments play only a subordinate role. It is also a process that goes on at a metropolitan, to some degree supra-metropolitan, scale.[8]

The mediation of actual instances of inward investment is a joint project of three types of agent: the gas and electric utilities, the local chambers of commerce, and the local governments.[9] Each agent performs a distinctive role in a shared division of labor. Initial contact by a firm considering a location in the area is almost invariably through the utilities. The local chambers provide more detailed information, particularly on labor matters to the firm in question and generally ease the way, perhaps troubleshooting with the local government. Local government brings up the rear, with decisions on incentive packages, including the extension of utility lines and tax abatements. This is the last stage in the process before a decision is actually made by the firm. The utilities dominate this process.

The utilities have been involved in local economic development for a long time. Every utility in the United States has its own economic development department and has since at least the 1930s. This is testimony to their dependence on revenue from a limited service area, to their investment in facilities of long life, and to the limits placed by very stringent takeover laws on spreading risks spatially. Owners of industrial sites typically lodge information on the sites they have available and what those sites offer, including infrastructural availability and locational attributes with these departments of economic development. As a result, utilities have very large data banks covering their service areas, areas that often go beyond defensible metropolitan boundaries. Given the propensity of site-seeking firms to spread their search over fairly wide areas—a reasonable strategy considering the high degree of specificity and limited substitutability of sites—the utilities are the logical first port of call.

Assuming that there is interest in at least one of the sites offered, the next stage in the location process is a site visit. The utility is likely to delegate much of this responsibility to a local chamber of commerce. Local chambers take a strong and enduring interest in local economic development. The chamber for a major city like Columbus or Cleveland will invariably have an economic development department as will many suburbs. Much impetus for this comes from property developers who have their own forms of local dependence: A local knowledge, a local reputation with builders and banks, and so forth make it hard for developers to operate outside particular metropolitan areas, so the efforts of the local chamber are of particular interest.

The chamber's role is a mixed one. For the site visit the chamber will put together a team of local businesspeople with whom representatives of the firm can meet. The chamber finds it easier to do this than the utilities would because existing businesses in the locality are not necessarily motivated to encourage new companies. New firms can pose threats to local labor markets and, in some instances, to product markets. The personal contacts of the chamber—perhaps a calling in of debts—can help. What the firm wants to learn from this team is largely a matter of labor: rates of absenteeism, rates of labor turnover, management-labor problems, the extent and character of union organizing, and the like. This knowledge is not likely to be found in the published statistics that the utility will already have compiled and made available.

Before this, the chamber may also have used its local knowledge to filter out certain sites. Sites in localities where there have been difficulties in the past with local residents are to be avoided. Later the chamber may be called on to pressure local government to put together various financial incentives and to act promptly—something that the utility finds it harder to do because it has no legitimate community role at that particular geographical scale.

Finally, there is local government. Much of the foregoing will usually have been carried on with local government having no knowledge—an interesting facet of the process that will be explained later, but that certainly underlines local government's subordinate role in what must be regarded as considerably important to the local government. The importance, of course, derives partly from the still substantial local tax base dependence of municipalities. Businesses pay property taxes and their employees increasingly pay local income taxes and, when spending their money in local stores, pay local sales taxes. In addition, elected officials depend on local sponsors for their campaign funds and developers are usually the first to contribute.

Once the site-searching firm has expressed strong interest in a site, the local government's task is to put together a package of incentives to clinch the deal. This can include extending water and sewer lines, pressuring the state to construct a new freeway interchange for very large investments, or even paying for labor recruitment and training. This will be in addition to the more common requests for rezonings and tax abatements.

All this means that the mediation of inward investment is far from straightforward. Several agents are involved in different ways. Diverse labors have to be integrated, and, as we will see later, market or quasi-market mechanisms are difficult to engage. In other words: Cooperation is necessary. This cannot be taken for granted. An inventory of the problems likely to arise includes at least three issues.

The first derives from the fact that localities of interest are unlikely to coincide. Utility service areas will include numerous municipalities. Different gas (or electric) utilities may divide large metropolitan areas between themselves. Chambers of commerce in those metropolitan areas may include some that are metropolitan in extent. In Central Ohio, the Greater Columbus Area Chamber has a membership coextensive with the metropolitan area as a whole. In its membership area, it coexists with much smaller suburban chambers. This can be a source of difficulty.

A utility with a lead obviously seeks location of the firm in question in its service area. A metropolitan chamber of commerce covering an area that is divided between two utilities, however, may not share that objective. Once a utility has the information, it may share it with the other utility, which could then approach the firm independently and even secure a location within that part of the metropolitan area that it services. Alternatively, the chamber can share the information with site owners who are chamber members but with sites outside of the utility's service area and to which they may then try to divert the firm's attention. I will call this the problem of territorial noncorrespondence.

A second problem is that of confidentiality. For diverse reasons, firms seeking sites usually want to keep information about their activities within a very tight circle until quite late in the location process. In an atmosphere of concern for plant closures, leakage of information about a new location can incite labor problems. On the other hand, it can also alert competing firms as to the nature of the firm's business strategy. The major difficulty here is likely to be elected local government officials. Publicity is their life blood. News of an imminent location of a major employer is seen as something that can raise their profile in a positive way. The temptations are great to make an announcement before the firm is ready for it.

A final area of difficulty faced by the utilities is eliciting more generalized cooperation from more local agents. The utilities have to be able to count on the ability of these local chambers and local governments to provide the necessary support in the attempt to convert prospects into actual investments. This support must include a realistic understanding of what is and what is not possible in the particular community, an ability to provide incentive packages, an ability to deliver what is promised, and an ability to troubleshoot objections from the local community. This problem stems mainly from the different incentive structures faced by the utilities and the more local agents. As long as the prospect locates in respective service areas, the utilities will gain. But given that it is habitual to provide the prospect with a choice of sites and given that the prospect may show interest in more than one of them, the same incentive for the locals may be lacking.

Primary among the social practices through which cooperation is secured, and these problems of integrating diverse labors overcome is that of relationships of trust. In the fieldwork on which this discussion is based this was emphasized repeatedly:

> Trust is another key word in this. [X] emphasized cooperation and our ability,
> his organization [the chamber] and ours [the utility] to be effective on behalf

of this region, [but] you [also] have to have a trust level at the local level with all these different entities. (Utility official, February 1993)

One of the things I have found and it goes back to . . . the networking in economic development, is the personal relationships that you've developed and the trust that you've developed with these individuals is 75 to 80% of it. (City official, November 1992)

That [lead information] comes with trust. Those kinds of new injections of information certainly come from developing the trust that when they tell us something that we're gonna hold it confidential. (Chamber official, November 1992)

A more specific norm operative here is what is known as "honoring the source." If information about a possible lead comes from the utility, no chamber or local government official will divulge it to anyone who might have an interest in trying to persuade the firm to locate outside of the utility's service area. This also extends further up in the chain of information. If the lead comes originally from the state, then no utility, even if they operate, as many do, in more than one state, will try to use the information to the benefit of an out-of-state site.[10]

Serious problems of maintaining confidentiality attach to local government. But that does not mean that some local government officials, particularly if they hold nonelective offices, cannot be part of the network of trust. Much depends on the particular individuals and their reputations in maintaining the confidence of the utilities and the chambers. To the extent that the chain of trust is broken somewhere, sanctions are used to reassert the need for cooperation. The failure of local governments to provide the necessary financial incentives can make the utilities unwilling to bring leads in their direction until there is a change of course:

The problem in [City X] is with the public school board. Everything's a political football over there. We had companies that came in here, good companies, that had to go through an inquisition over there. . . . They made them look like a war criminal. . . . We had a guy go in there who wanted to create about 200 jobs in [the city] and put about a $75 million investment in [the city] who said, "They made me feel like a Chicago mobster." . . . We used to avoid the City of [X]. . . . There were some people in this business who said I'm not going to mess with them, I'm done with them, the heck with them. (Utility official, June 1992)

The role of trust as a primary mechanism of cooperation in this instance can be understood in terms of our earlier, albeit brief, discussion. Where transactions are possible with a range of substitutable alternatives, that possibility disciplines those involved in the exchange and limits opportunism and malfeasance. But in the particular instance being discussed here, substitutability is quite obviously limited and in several distinct directions.

From the utility's standpoint, a major problem is that of sunk investments. Every project to bring in a new industrial investment is different. Time and energy are devoted to putting together a marketing job oriented to a particular site in a particular place and requiring the orchestration of the activities of particular individuals. Should the project fail, there is no way in which the investment that has been made can be transferred to another marketing project oriented toward the almost certainly very different needs of another firm. The failure of local governments and local chambers to cooperate in bringing about the successful location of the firm, therefore, can be particularly damaging. The specific needs of the site-seeking firm, moreover, mean that the number of sites that the utility can offer is going to be very limited. This reduces the utility's power vis-à-vis the more local agents even further because the degree to which agents can be played off against one another is limited.

Conversely, the problem from the standpoint of these more local agents is the monopoly that the utilities hold over the supply of lead information. Utilities have monopolies within respective service areas. Their inventories of sites make them the gatekeepers as far as site-seeking firms are concerned. Going to another utility for lead information will not help a locality because it will not be in the area served by that utility. This, of course, gives the utilities the power to enforce confidentiality, norms of honoring the source, and other forms of cooperation, as I noted in the discussion of tax abatements in City (X).

## Conclusion

Local governments have major interests in local economic development. To secure their own respective tax bases, they have to attract in new investment in the form of real estate developments, corporate headquarters, branch plants, and the like. What local government has to offer is important to those same capitalist interests. This is the basis for the insight in urban regime theory that local governments need to construct coalitions and other forms of cooperation with these interests. This has led to studies exploring the nature of the subsequent politics in different cities and classifying the sorts of regimes that result from them.

The nature of these cooperative arrangements, however, is much more contingent than seems to be allowed by urban regime theory. The mix of public and private involvement, whether or not organization has to be extended across local government boundaries, is highly variable. A major part of the problem, in other words, has been the failure to theorize fundamental interests in local economic development and to separate them from the (contingent) conditions of their realization. The case studies in this chapter have shown just how variable they can be.

## Notes

1. See Jessop (1995b) for a good discussion of the relations between the governance literature and regulation theory.

2. In a now classic article, Davis and Whinston (1965) laid out the conditions for state intervention into the rehabilitation of housing. These conditions included market failures resulting from externalities and the obstacles to land assembly erected by the owners of strategically placed land. In this context, state intervention aimed at internalizing these externalities and facilitates land assembly through the right of eminent domain would encourage investment in housing.

3. Implicit in my discussion throughout this chapter is a particular conception of "local economic development." This conception is hegemonic for the politics of local economic development in the United States: Local economic development occurs through inward investment. Toward that end local governments make their own investments, primarily in physical infrastructure, and lobby the state for changes in labor law and state taxation, which will enhance the local business climate. In contrast with other countries, therefore, this conception does not emphasize building up infrastructures of a more social kind. This concept is, however, implicit in the urban regime literature, a literature that has for the most part taken as its point of reference the American case.

4. The city departments involved have included not just the public utilities division providing the water and sewer services but also police and fire. Their acquiescence is necessary because of the increased service load subsequent to annexation.

5. For the City of Columbus, the rationale for the annexation policy has always been fiscal: to capture suburbanizing land uses for the city's tax base. Harrison Smith, a land use lawyer who has mediated numerous annexations to the City of Columbus and who is a principal ideologist for that policy, has put it aptly: "In a static central city where there remains for the resident, the retailer, or the office tenant a continuing option to move out of the city, the effect of time is devastating. Everything gets old. Maintenance costs go up, rents go down, and municipal costs go up. The process accelerates and the result is—Cleveland" ("Columbus Annexation Policy Pays Off," 1979).

6. For a more detailed discussion of the issues discussed in this case study see Cox and Jonas (1993).

7. In 1985, of more than 1,800 building permits issued in Columbus for single-family homes, less than 1% were for homes in Columbus School District ("Home Builders Fear Annexation," 1986).

8. This section draws significantly on Wood (1993).

9. I am confining my attention here to the mediation of industrial investments. The case of commercial investments is different: the utilities do not play a leading role in that instance.

10. Cincinnati Gas and Electric operates both in southwestern Ohio and in northern Kentucky. To avoid any possibility of failure to honor the source, the utility has two economic development departments, one for each state.

# 7

# Regulation, Regime, and Practice in Urban Politics

JOE PAINTER

In this chapter, I want to examine the relationship between urban regime theory and regulation theory in terms of their *theoretical commensurability*. However attractive the integration of the two perspectives may appear, it can only provide a sound framework for understanding urban politics if there is at least no contradiction between the conceptual bases of the different elements. As regulation theory has many versions, I begin by outlining briefly my understanding of regulation theory and of its relationship to urban politics. I go on to suggest that this account, although helpful, has inherent limits, which derive from the methodology of regulation theory itself. One question addressed by this book is whether urban regime theory can be linked to regulation theory to allow these limits to be transcended and thereby to enable urban theorists to produce stronger, more powerful or more complete analyses of urban politics. Although much in regime theory is commendable, some conceptual ambiguities challenge its compatibility with the regulation approach. I therefore conclude by proposing a reworked formulation of the regime idea for interpreting urban politics. In place of the rational choice model of political practice that underlies Clarence Stone's (1989) formulation of the concept of urban regimes, I propose an approach based on a critical engagement with aspects of the work of Pierre Bourdieu. This focuses on the relationship between political practice and processes and sites of regulation and counterregulation.[1]

## Regulation Theory and Urban Politics

Together with colleagues, I have spelled out an approach to regulation theory and its relationship to urban politics in detail elsewhere (Bakshi, Goodwin, Painter, & Southern, 1995; Goodwin & Painter, in press; Painter, 1991, 1995; Painter & Goodwin, 1995) and there is space here for only a brief summary. Regulation theory starts from the premise that complex social systems are conflictual, contradictory, and crisis prone. These characteristics generate tendencies to failure of system reproduction and thus to system breakdown and collapse. Historically, however, although such failures and breakdowns do occasionally occur, there are numerous instances of successful system reproduction despite these countervailing crisis tendencies. Systems in which crisis tendencies are mitigated in this way are said to be "regulated." The process of regulation is nonnecessary; if it were guaranteed by the operation of the system's core features the system would be self-regulating. In contrast, regulation arises contingently, and is not necessarily, or even usually, the intended result of a deliberate strategy, though it frequently *can* be explained as the product of the interaction of the unintended consequences of intentional actions.

This formal account of regulation tells us nothing about the social content of the systems being regulated. In practice most regulation theory has taken the object of regulation to be the economy and, in particular, the process of capital accumulation within industrial capitalism. There is no intrinsic reason why the approach could not also be applied to other complex systems such as the state, or an urban political system, however. For now I shall concentrate on the case of the capitalist economy. Regulation theorists argue that the process of capital accumulation is reproduced at any given moment through the dynamic of the prevailing *regime of accumulation*. A regime of accumulation specifies the broad relationships between production, consumption, saving, and investment, and also defines the geographical extent and degree of autonomy of the circuit of capital (national, continental, global, etc.). A fundamental insight of regulation theory is that the history of capitalism has been marked by a succession of qualitatively different regimes of accumulation, each defining a different set of relationships between production, consumption, saving, and investment. Regulation theory shares with classical Marxism a recognition that abstract features of the accumulation process are invariant over time (such as the existence of the wage relation) but places much more emphasis than did classical Marxism on the historically and geographically variable forms that those features take as the process of accumulation is revolutionized through successive phases of struggle, crisis, and restructuring.

Although the prevailing regime of accumulation describes the dynamic links between different elements of the accumulation process, it cannot guarantee their reproducibility over time and space. Indeed, the inherent crisis tendencies of the capitalist mode of production are expressed in the regime of accumulation as, for example, periods of underconsumption or, conversely, underinvestment. According to regulation theory, if the regime of accumulation survives (and that is not inevitable) it does so because the relationships between its elements are being *regulated*. Regulation theory regards regulation as a complex, uneven, and contingent (rather than functionally necessary) process. The mechanisms of regulation vary over time and across space but commonly include institutional forms, social relations in civil society, and cultural norms as well as those activities of the state and judicial systems that the term *regulation* more commonly connotes in Anglophone writing. It seems that the effectiveness of regulation often depends on, or is enhanced by, the contingent interaction of these mechanisms.

In conventional regulation theory it is often argued that regulatory mechanisms combine in integrated, relatively stable, and relatively coherent *modes of regulation*. In political terms, modes of regulation have sometimes been presented as historical grand compromises between capital and labor. The concept of mode of regulation has been criticized for tending to overemphasize stability at the expense of change, compromise at the expense of struggle, and structure at the expense of agency. For these reasons, as Mark Goodwin and I (in press) have argued elsewhere, it may be preferable to think of an historically variable *process of regulation* rather than a succession of discrete "modes" of regulation. This would admit the key regulationist points that cultural and institutional influences on accumulation are centrally important and that the nature and intensity of conflict vary, while recognizing that regulation is itself both the medium, object, and outcome of social struggles and conflicts and subject to crisis and restructuring.

Many regulation theorists identify the 30 years after the end of the Second World War as a period of particularly successful regulation in many industrialized capitalist countries. During this time, a regime of accumulation based on a link between mass production and mass consumption was made possible by regulation that drew together (among other things) (a) state involvement in setting minimum living standards, (b) institutionalized collective bargaining to give workers a share of productivity gains, and (c) a "national" circuit of capital linked to national systems of monetary and financial regulation and an international system of stable exchange rates. Conventionally, regulation theorists have regarded these relationships as constituting a mode of regulation and have labeled it fordism.[2] With the breakdown of fordism from the mid-1970s onward, the regulatory process has undergone significant restructuring. Commentators disagree about the impact of these changes. For some they herald the emergence a fully fledged postfordist mode of regulation constituted around customized production, niche (rather than mass) consumption, flexible wage bargaining, and financial deregulation and globalization. Others accept that a new mode of regulation is a possibility but argue either that its development to date is at best embryonic (Jessop, 1992a) or that there are a variety of alternative feasible postfordisms. Another perspective uses the relative indeterminacy of future forms to argue in a normative fashion for one future "compromise" rather than another (Lipietz, 1992). Finally, some commentators argue that the present period is marked by the continuing turmoil of the breakdown of fordist forms and its aftermath, with little evidence that a new mode of regulation is emerging, or, under present conditions, could emerge (Peck & Tickell, 1994).

As well as differentiating capitalism through time, the concepts of regime of accumulation and mode of regulation can differentiate it across space, and partly for this reason geographers and urban and regional theorists have been particularly interested in the regulation approach. With its focus on the institutional and cultural influences on economic growth and development, regulation theory provides a way of understanding the spatially uneven character of economic change, at least in principle. In practice, regulation is often tacitly assumed to operate at the scale of the nation state. This makes some sense when we consider fordism (though the extent to which fordist regulation was socially and spatially differentiated within states has often been underestimated; Bakshi et al., 1995). There is no reason to suppose, however, that modes of regulation *must* be secured at the level of the nation-state, still less that the ongoing process of regulation (whether it constitutes a coherent mode or not) will be national.

Their concern with uneven development has led urban and regional researchers to propose different ways in which theories of regulation might be spatialized. Among others, Esser and Hirsch (1989) examined the impact of the supposed emergence of postfordism on urban and regional systems; Florida and Jonas (1991) discussed the connections between urban policy and modes of regulation; Jessop (1994) argued that the postfordist state will typically be hollowed out with functions shed downward to regional and local government and upward to international structures like the European Union; Peck and Tickell (1992) have discussed the concept of local modes of social regulation; Mayer

(1994) argued that postfordism will provide new opportunities in the practice of urban politics; and my colleagues and I proposed the concept of local regulatory capacity (Painter, Wood, & Goodwin, 1995).

These debates suggest that the issue of geographical scale is not in itself an insuperable problem of the regulation approach.[3] This means that any claim that regulation theory needs urban regime theory *because it deals with the urban scale* is misguided. It is untrue to say that regulation theory necessarily deals with large scale (regional, national, or international) and that regime theory is needed to fill in the gap at the urban or local scale. It is also fallacious to equate on the one hand macrolevel theories with abstraction and large geographical scales or on the other hand meso- and microlevel theories with successively more concrete analyses and successively smaller geographical scales. One can have abstract theories of micro processes, mesolevel theory does not apply exclusively to "middling" geographical scales, and concrete accounts can be written of large-scale processes. Therefore even if regulation theory is a macrolevel theory, that does not mean that it needs a mesolevel theory *to deal with the urban scale* (though it might need one for other reasons). Nevertheless there *are* limits to the usefulness of regulation theory. These derive not from the problems of addressing geographical scale but from its approach to explanation. If we are to argue that regime theory is a necessary complement to regulation theory, we must do so on the basis that it adopts an approach to explanation that is both compatible with regulation theory and that provides explanations of phenomena for which regulation theory cannot account.

# The Limits of
# Regulation Theory

The debates around uneven development, scales of regulation, and the impact of regulation on urban and regional restructuring (and vice versa) represent a new level of sophistication in the development of the regulation approach. As the debates proceed, however, it is becoming clear that they are also revealing some limits of the regulationist framework as a method of analysis. The existence of such limits should not be taken to mean that the regulation approach is somehow "wrong," rather, there are some questions that it does not (or is unable to) address.

In crude terms, the regulation approach explains (certain characteristics of) the object of regulation (usually the development of capitalism) in terms of the process or mode of regulation (institutional forms, cultural norms, state structures). Actually, the explanation is more complex than this. Objects of regulation do not precede regulation, existing in some preregulatory limbo awaiting the emergence of a mode of regulation. Rather, processes and objects of regulation emerge together and are produced by one another. Nevertheless, the explanandum of conventional regulation theory is the process of capital accumulation, whereas its explanans is the complex of social, political, and cultural processes and practices that sustain accumulation in the face of its tendency toward crisis and collapse.

Within this explanatory schema, urban politics features (if at all) as a possible element in the explanans.[4] Regulation theory as written hitherto explains economic continuity and change in terms of (among other things) political processes. It is not an explanation of those political processes. This distinction is essential to avoid a slide into functionalism. The processes of regulation are identified by their effects on the accumulation of capital. To argue that certain contemporary changes in institutions, political practices, or cultural norms are accounted for by a shift from (for example) a fordist to a postfordist mode of regulation is to explain change by its effects (because it is the new mode of regulation that is produced by the changes, rather than vice versa).[5] Such an approach would be functionalist and thus fallacious.

Conventional regulation theory is good at explaining the dynamic of regulatory processes once they are established and the crisis tendencies that undermine regulation and produce regulatory breakdown and failure. Because modes of regulation are understood to be the product of the interaction of contingent phenomena, the concept of mode of regulation cannot explain the emergence of those phenomena in the first place. As Jessop (1990b) argues, "unless one examines the mediation of regulation in and through specific social practices and forces, regulation will either go unexplained or will be explained in terms of 'speculative' structural categories" (p. 319). This mediation is the focus of urban regime theory.

Though the form and nature of urban politics cannot be unproblematically derived from the characteristics of the prevailing mode of regulation, urban politics is not straightforwardly independent of the mode of regulation either— in practice they are partly mutually constituting. What is required is an approach to the analysis of urban politics that can unravel the causal processes that explain it whether they are grounded in the mode of regulation, in practices that are counterregulatory, or in other spheres of social life that have no strong relationship to the regulation of capital accumulation at all.

The explanatory power of regulation theory is therefore limited. The notion of regulation may be a necessary component of a satisfactory account of urban politics, but it is far from a sufficient one.

# Urban Regime Theory

The concept of *urban regime* has gained significant prominence in the literature of urban studies and political science, especially in North America (Elkin, 1987; Fainstein & Fainstein, 1983; Lauria, 1994a, 1994b; Orr & Stoker, 1994; Stone 1989, 1993). Recently it has been gaining popularity in relation to British urban politics (DiGaetano & Klemanski, 1993b; Lawless, 1994). In addition, Stoker and Mossberger (1994) have argued that it is possible to strip regime theory of some of its ethnocentrism and use it more widely, including in comparative studies (see also DiGaetano & Klemanski, 1993a; Harding, 1994; Stoker, 1995). Within this developing literature the idea has been formulated in most detail by Clarence Stone, especially in his pioneering study of Atlanta (Stone, 1989). The term has been widely adopted, however, and has sometimes been used in rather imprecise ways; as Stoker puts it, "Regime terminology is used, but a regime analysis is not really provided" (1995, p. 55).

Regime theory starts from the proposition, which it shares with regulation theory, that the process of governance in complex societies is about much more than government. Successful governance, whether of a city, a nation-state, international relations, or economic processes almost always depends on the availability and mobilization of resources and actors beyond those that are formally part of government. Governing a city, particularly in the United States where the institutions of elected urban government are relatively weak, relies on the ability to form governing coalitions that bring together the formal agencies of government with interest groups from the wider society. Foremost among these, in the American context at least, are business interests. Stone (1989) carefully points out, however, that although the prominence of the business connection is hardly surprising, a regime is not inherently a coalition with business (or with any particular interest, come to that):

> In defining an urban regime as the informal arrangements through which public bodies and private interests function together to make and carry out governing decisions, bear in mind that I did not specify that the private interests are business interests. Indeed in practice, private interests are not confined to business figures. Labor-union officials, party functionaries, officers in nonprofit organizations or foundations, and church leaders may also be involved. (p. 7)

On the other hand, business interests are central in practice because regime success is evaluated (at least in part) by economic prosperity and because (in the United States but not, for example, in Britain) local governments depend heavily on local businesses for tax revenues. Outside the United States, other interests may be more central to the coalition. In many European cities, for example, appointed local state officials, technocratic managers, and professionals play a central role. Furthermore, in many cases the most important business interests are not locally embedded to the extent evident in many examples discussed in the literature on the United States.

An urban regime can thus be defined as a coalition of interests at the urban scale, including, but not limited to, elected local government officials, that coordinates resources and thus generates governing capacity. The notion of governing capacity is important in Stone's account because it relates to his conception of power, which, he argues, is distinctive. Unlike conventional conceptions of power, which emphasize social control, or "power over," Stone claims to work with a social production model of power that emphasizes "power to." Governing capacity, or the capacity to act, is produced through coalition building. There is also an assumption, not always explicit, that a coalition must endure through time to qualify as a regime. A group that comes together to pursue a particular project and then dissipates again is not a regime. One hallmark of a regime is the willingness of actors to maintain membership of the coalition even when it is working against their short-term interests. In Stone's Atlanta case study, for example, despite the segregationist views of many of its members the downtown business elite was prepared under certain circumstances to accommodate the interests of the black middle class to maintain access to political power. For Stone, the governing coalition constitutes a single regime for as long as its members (or the interests represented in it) remain the same.

From Stone's account, it might be expected that an urban regime is rather an unusual phenomenon. He clearly regards the stability and strength of the Atlanta regime as in some ways remarkable and frequently contrasts the relatively peaceful community relations across the racial divide in Atlanta with the much more turbulent histories presented by other Southern cities. One outcome of this (and measure of regime success) is the strength of the Atlantan economy. On Stone's definition, cities without strong coalitions and marked regime effects should surely be regarded as nonregime cities. Elsewhere in the literature, however, there is an assumption (albeit sometimes an implicit one) that all, or at least most, cities have regimes and that one task of the urban political scientist is to categorize regimes into different types. Stoker and Mossberger, for example, identified three broad regime types, each with subcategories. Stone himself developed a fourfold typology of urban regimes that encompasses maintenance regimes, development regimes, middle-class progressive regimes, and regimes devoted to lower-class opportunity expansion (Stone, 1993). The inclusion of maintenance regimes in particular seems to sit oddly with the definitions advanced in the Atlanta study. If a regime involves building a coalition between the local state and private interests to generate governing capacity, in what sense is a maintenance regime (which seems to require few resources, limited involvement of nongovernmental interests, and little active governing) a regime? This broadening of the regime concept is clearly intended to enable its application in a wider range of cases and contexts, yet it runs the risk of blunting the definition of the original formulation and, in some cases, simply using the term *urban regime* as a synonym for "urban politics" or "urban governance."

Conversely, just as I have argued that the methodology of regulation theory would not be invalidated by the absence of empirical examples of modes of regulation it could be suggested that the regime approach is not invalidated by the relative rarity of successful, enduring regimes. First, the concept of urban regime has some counterfactual force, defining what kinds of alliances would be required if certain outcomes were to be produced. Second, just as no economic activity can continue for long without a certain amount of regulation, so it might be argued no city can maintain even limited social and economic coherence without some governance, which is in turn the product of and dependent on the bringing together of a variety of agents in some form of a regime. From this perspective, a regime need not be very successful or very long lasting to be called a regime. As a coalition of social forces, a regime strives to govern, but that does not guarantee successful governance.[6] This kind of argument provides a starting point for extending the regime concept to give it more general applicability without at the same time producing insuperable problems of empirical validation.

If regimes are understood as dynamic forms that are in a continual process of formation and becoming while facing challenges and countervailing pressures, then there are some strong formal similarities with the reworked notion of regulation that I sketched above and have outlined in detail with Mark Goodwin elsewhere (Goodwin & Painter, in press). Just as I have argued that regulation theory should emphasize regulation as a process, rather than as an established state, so the emphasis in regime theory could be placed on understanding the processes and struggles involved in regime formation, reproduction, and crisis. Such an emphasis holds out the prospect of a regime theory that is theoretically commensurable with regulation theory. Whether the prospect is realized, however, will depend on precisely how the social and political processes that go into building, maintaining, or undermining a regime are conceptualized. In Stone's account, the process of regime formation is understood as grounded in rational strategies pursued by political actors. In what follows, I suggest that this is an inadequate characterization of political process but that a regime theory that problematizes the notion of rational strategy by, for example, drawing on Bourdieu's concept of *habitus* has the potential to provide some elements missing from the regulationist account in ways that are more theoretically compatible.

## Conceptual Ambiguities in Regime Theory

The apparent appeal of the urban regime concept for a regulation theorist interested in urban politics lies in the emphasis in regime theory on the political processes at the urban scale, combined with its rejection of conventional pluralist and elitist approaches. On the face of it, regime theory is well placed to provide some of the explanatory links missing in the regulationist account, in that it focuses explicitly on the content of political disputes and on forms of political conflict and cooperation at the urban scale. As I showed earlier, the attraction of regime theory stems from its focus on politics and not from its concern with the urban scale. Potentially, therefore, regime theory might help explain the emergence (or, indeed, the nonemergence) of regulatory practices and thereby provide an account of the emergence and consolidation of regulatory processes at the urban scale. Despite its apparent attractions, however, several ambiguities associated with the regime approach, as presently formulated, question its commensurability with regulation theory.

The first of these relates, paradoxically, to a formal similarity between the regulationist and regime approaches. I argued earlier that regulation theory explains economic growth and development in terms of the operation of political processes, among other things. At one level, clearly, this is also exactly what regime theory does. The economic growth pattern of Atlanta in the postwar period is in part explicable, according to Stone, as an effect of the biracial urban regime in the city. It is tempting, therefore, to regard urban regime theory as a kind of regulation theory of the urban, with the regime representing a sort of urban mode of regulation. Unfortunately, however tempting, this maneuver would, I think, be misguided.

First, although urban regimes might well have certain regulatory effects, the process of regulation is significantly more complex than the operation of a regime. Many key components of regulation, even at the urban scale (such as the social organization of labor markets) fall outside the scope of a regime. Although a regime might be part of the process of regulation, it cannot form an urban mode of regulation on its own. Second, the idea of an urban (or local) mode of regulation is in any case deeply problematic because the processes of regulation affecting economic activity in a given urban area may well lie outside of the area. The regulation of local economies is not necessarily exclusively, or even mainly, a local matter. Third, and most important for my argument, identifying an urban mode of regulation would not in any case solve the problems posed by the explanatory lacunae in the regulation approach and spelled out above: Namely, how do we account for the emergence of regulatory processes in the first place? To pose the question more directly, the existence of an urban regime can help to explain urban economic growth, but what explains the existence of an urban regime?

Here another conceptual problem arises—one which, in my view, compromises the usefulness of Stone's arguments from a regulationist point of view. Unlike regulation theory, regime theory *does* supply an explanation of the political process. In Stone's account, the explanation is not fully developed in detail, and some of its assumptions are implicit rather than clearly stated. Although this leaves some ambiguity around the explanatory framework being adopted, my provisional conclusion is that this framework is not compatible with the regulation approach.

The explanation of the regime phenomenon advanced by Stone has two main components. The first of these, it seems to me, *is* compatible with regulation theory. In contrast with the pluralist approach, Stone argues that different groups in the city have differential access to regime membership and that these differences are the product of structural inequalities in the distribution of resources. (Note that "resources" does not mean material resources only but could include cognitive, social, and symbolic resources.) Business elites control resources that make them both more attractive to local governments as coalition partners and better placed than less resource-rich groups to negotiate regime membership.

Access to resources thus allocates certain groups the status of potential regime members. The second component concerns the process through which potential regime members become and remain actual (and active) coalition partners, and here I find Stone's (1989) argument flawed. According to Stone, the emergence, nature, and reproduction of an urban regime is explicable in terms of selective incentives (pp. 186-191). The argument runs as follows: Successful urban regimes involve cooperation between partners whose immediate interests do not coincide directly and that can even be opposed. Although it may well be understood that there are long-term benefits to cooperation that outweigh the costs of cooperating, those benefits are likely to accrue to all potential regime members, whether they play a part in the regime or not. (Business-friendly policies help businesses that are not directly involved in the governing process as well as those that are.) Because cooperation involves the expenditure of time and effort, and the subordination of immediate interests to long-term and possibly rather uncertain future gains, individuals have little incentive to cooperate, especially if any longer-term benefits that are produced are likely to be distributed widely. In short, the major problem facing a would-be regime builder is how to prevent potential regime members from gaining the benefits of regime participation without incurring any of the costs. As Stone points out, this is the classic free rider problem familiar to theorists of collective action (Olson, 1965).

Stone's solution to this conundrum is to propose that regime involvement is governed by a system of selective incentives. Selective incentives resolve the free rider problem by offering additional benefits to those potential cooperators who do in fact cooperate and denying benefits to those who do not:

> The traditional solution to the collective-action problem has been selective incentives; that is, to supplement group benefits by a system of individual rewards and punishments administered so as to support group aims. Those who go along with the group by paying dues, respecting picket lines, and so on, receive individual rewards and services; those who do not lose valuable benefits or incur sanctions. Voluntary efforts are thus complemented by inducements or coercion, individually applied. (Stone, 1989, p. 186)

Stone points out that the need for selective incentives varies according to coalition goals. Coalitions that pursue large-scale, resource-hungry, and risky projects will have greater need of additional incentives to encourage coopera-tion. In contrast, "caretaker" regimes will not need to mobilize collective endeavor to the same degree and will have less need of selective incentives.

Although the concept of selective incentives is not the focus of extended discussion in the Atlanta study, it is crucially important because it is, for Stone (1989), at the core of why cooperation (and hence, regimes) occur at all. Selective incentives are "what holds a governing coalition together" (p. 175), and although he recognizes that there are other factors at work, "control of selective incentives is a significant factor in determining which alignment of groups will be best able to press its case as the community's governing coalition" (p. 190).

The use of the selective incentives concept as the core of the explanation of regime origins and reproduction means that, as an explanatory framework, regime theory is grounded in the methodology of rational choice theory. Within the regime perspective, the political process is understood (in large part, at least) in terms of decision making in the face of patterns of costs and benefits in which means-end rationality is deployed to provide the greatest returns to self-interested individuals. Where "unusual" outcomes are observed, such as sus-tained cooperation across apparently deep social divides, Stone appeals to recent developments in game theory (Axelrod, 1984) to show that such outcomes are explicable in rational choice terms.

If the concept of urban regimes is ultimately grounded in the rational choice model, it is difficult to see how it can be commensurable with regulation theory. Regulation theory precisely rejects the idea that processes of regulation arise through individual choices governed by the calculation of rational self-interest; yet, as I have suggested, neither are they explicable by functional necessity at the level of the system. What is required is a concept of political practice and strategy that can inform an analysis of urban politics without contradicting the methodological stance of regulation theory. If urban regime theory is to provide the complementary account of (political) agency that regulation theory currently lacks, then it will require substantial reworking to eliminate its rationalist and individualist connotations.[7]

## Toward a Theory of Practice in Urban Governance

In the remainder of this chapter I want to explore the scope for just such a reworking provided by a selective appropriation of ideas from the work of Pierre Bourdieu, and especially from his discussion of *The Logic of Practice* (1990). Bourdieu's writings are rich and complex, but his writing style is far from straightforward. This, coupled with a perception that his work is concerned mainly with cultural practices and is associated with anthropology and sociology, means that his ideas have gained little currency in writings on the state, urban politics, and political theory. Let me say at once that I am not proposing that Bourdieu's ideas can resolve all the tensions between regulation and regime approaches; nor are his ideas themselves without their problems (Calhoun, LiPuma, & Postone, 1993). Nevertheless, if processes of regime formation, consolidation, and crisis are to be used effectively in explaining the character of urban politics and of its contribution or otherwise to effective regulation, then it is crucial that our investigations of such processes are informed by an adequate understanding of political practices. Although Bourdieu does not address issues of political practice explicitly, the idea of practice itself is central to his work, and is formulated in terms that explicitly reject rational choice perspectives.[8]

Bourdieu argues that conventional theoretical approaches can be divided into two broad categories: objectivism and subjectivism. Neither of these, he suggests, is capable of accounting adequately for practice. This is hardly a novel argument, of course, but transcending the division between objectivism and subjectivism seems extraordinarily hard. For Bourdieu, objectivism is epitomized, but not limited to, structural anthropology. In its concern only with the objective conditions of practice, objectivism is unable to account for the relationship between those structures and the experiential meanings that are implicated in practice. Paradoxically, Bourdieu (1990) writes of objectivism, "beneath its air of radical materialism, this philosophy of nature . . . amounts to a form of idealism" (p. 41). In contrast, subjectivism suffers from the inverse paradox. For Bourdieu, the paradigm case (but again not the only example) of subjectivism is the Rational Actor Model. As I have suggested, the rational actor model underpins the notion of political practice implicit in conventional regime theory. This approach locates the origins of practice in the mental decision making of the rational actor, which appears to constitute a voluntaristic approach to the explanation of practice. As Bourdieu points out, however—and this is the mirror-paradox to that associated with objectivism—by proposing that practice is governed by the rational calculation of self interest, the rational actor model in fact involves a determinism in which practice is governed by the objective conditions defining an actor's interests.

For these apparently opposed positions, which paradoxically appear to collapse into each other on closer inspection, Bourdieu would substitute a theory of practice. Practice for Bourdieu (1990) is neither the determinate outcome of objective structures nor the product of voluntaristic decision making:

> One has to escape from the realism of the structure, to which objectivism
> necessarily leads when it hypostatizes these relations by treating them as
> realities already constituted outside of the history of the group—without
> falling back into subjectivism, which is quite incapable of giving an account
> of the necessity of the social world. To do this, one has to return to practice,
> the site of the dialectic of the *opus operatum* and the *modus operandi* ; of the
> objectified products and the incorporated products of historical practice; of
> structures and *habitus*. (p. 52)

Thus, Bourdieu has no wish to dispense with the concept of structure—on the contrary, it remains central. Rather than taking structure as pregiven, however, he problematizes its conditions of production and considers how structures are themselves produced through social practice. Equally, he does not dispense with categories such as experience or subjectivity but, again, transforms them from primordial or essential features into social products that have their own conditions of existence and processes of determination.

The concept of *habitus* is central to Bourdieu's (1990) theory of practice. In a passage that hints at complementarity with the regulation approach, it is defined as

> systems of durable, transposable dispositions, structured structures predis-
> posed to function as structuring structures, that is as principles which generate
> and organize practices and representations that can be objectively adapted to
> their outcomes without presupposing a conscious aiming at ends or an express
> mastery of the operations necessary in order to attain them. Objectively
> "regulated" and "regular" without being in any way the product of obedience
> to rules, they can be collectively orchestrated without being the product of
> the organizing action of a conductor. (p. 53)

For Bourdieu, the concept of habitus is the mediating concept between structures and practices (Thrift, 1983, p. 30). Although the concept of habitus is somewhat opaque, it is clearly not a synonym for context. Bourdieu (1990) opposes those who would "correct the structuralist model by appealing to 'context' or 'situation' to account for variations, exceptions, and accidents . . . situational analysis remains locked into the framework of rule and exception" (p. 53). The implications of Bourdieu's formulations are that all practice is generated through habitus. Habitus consists of *dispositions,* a term that in French carries the twin meanings of that which disposes one to act in a certain way (predisposition) and that which is the result of a process (arrangement or distribution). Habitus is thus both product and generator of practice, but in generating practice, it predisposes rather than determines. In much of Bourdieu's (1990) work, the notion used in analysis tends to be class habitus, but there is no reason in principle why relations other than class should not be generative of habitus. Habitus has an infinite capacity for generating "thoughts, perceptions, expressions and actions" but only those compatible with its own conditions of production:

> The most improbable practices are therefore excluded, as unthinkable, by a
> kind of immediate submission to order that inclines agents to make a virtue
> of necessity, that is, to refuse what is anyway denied and to will the inevitable.
> (p. 54)

According to Bourdieu, the notion of habitus operates in relation to the concept of *field.* The field delineates the scope of operation of habitus by differentiating that part of the social whole in which the practical sense involved in the operation of habitus is effective. Beyond the field lie other fields in which the rules of the game, and hence the habitus, are different.

There is space here for only the most preliminary outline of the relationship between the concepts of practical sense, habitus, and field and the idea of urban regimes, and this is not an attempt to present a developed Bourdieusian urban political theory. In any case, as one commentator puts it, the usefulness of Bourdieu's work lies in its being "good to think with" (Jenkins, 1992) rather than a template that can be unproblematically laid over any area of substantive investigation.

As I showed earlier, what regulation theory needs, and what regime theory purports to offer, is a theoretically informed account of the dynamics of the urban governance process. Regulation theory is well designed to interpret the impact and effects of regimes on the urban economy because the whole approach incorporates the idea that economic activity is socially and politically mediated and produced. The regulationist account is weak in explaining the genesis and formation of regimes (and other political processes) in the first place. At the same time, Stone's use of the rational choice concept of selective incentives is not compatible with the regulationist perspective.

The notion of habitus provides an alternative approach[9] to understanding the processes by which potential participants in a regime come to join the coalition (or not, as the case may be). Focusing on the ways in which particular groups of actors make practical sense of their political world it problematizes the idea of rational decision making. First, habitus does not start from the erroneous assumption that political decisions are rational, or rationally arrived at. Second, and by extension, habitus allows a whole range of other influences to be brought into an analysis of regime formation. Questions of bureaucratic culture, ethical judgment, irrational assumptions, trust and mutuality, local chauvinism, political ideology, and a host of others take their place as potential parts of a multicausal explanation of political behavior.

Centrally important in habitus formation is the role played by knowledge, information, and political socialization. Different regime participants do not just have different amounts of knowledge about the conditions under which they are acting, they also *know in different ways,* and these different ways of knowing bear heavily on decisions about whether or not to participate in any particular governing arrangement. In the habitus knowledge appears to be instinctive and natural: It is labeled common sense and determines the actors view of the field and of the prospects associated with particular courses of action. A focus on the production of particular types of knowledge and ways of knowing in different actor groups therefore provides a way of "operationalizing" the notion of habitus in research terms.

This can be illustrated with some (admittedly speculative) examples of the habitus of different actor groups in the field of urban governance, which are summarized in Table 7.1. Note that this is not intended as a comprehensive list of the actors involved in urban governance, the precise contents of which will vary according to the empirical case being considered.

This formulation, although merely an outline, can provide the starting point for an approach to urban regimes based on the following six propositions.

**Table 7.1** *Habitus* of Different Actor Groups in Urban Governance

| *Examples of Actor Groups* | *Habitus Grounded In* |
|---|---|
| Local politicians | Political socialization through past party or community activism or machine politics; common sense based on political deal making and "fixing" or grassroots support and legitimacy |
| Private sector business managers | Socialization associated with entrepreneurialism, being businesslike, "getting things done," ends justifying means, profitability, local embeddedness, business and property security; different values dominate in different sectors and vary with firm size and ownership relations |
| Public sector professionals (e.g., urban planners, accountants) | Codes of professional conduct; procedures heavily influenced by norms and expectations generated in the process of professional training and accreditation; common sense based on detachment, objectivity, and public service |
| Public sector bureaucrats | Bureaucratic knowledges; norms associated with accountability, hierarchy, record keeping, and surveillance; means predominate over ends; maintenance of organizational structure |
| Unelected public bodies (e.g., U.K. local Quangos) | Knowledges imported (mainly) from private sector, though with some public service elements; ends predominate over means; culture of confidentiality; culture of formal, legal (rather than democratic) accountability; common sense based on getting the job done |
| Community organizations | Quite variable but can include knowledges based in combination of concrete experience and abstract ideals; common sense frequently based on "us-them" or "David-and-Goliath" metaphors; cultures of self-help coupled to rhetorics of civil and social rights |
| Voluntary sector | Varied; frequently grounded in notions of charity or self-help; rhetoric of "serving the community"; knowledge base varies with size and type of organization from amateurism and "muddling through" to highly professionalized |

1. There is no one unitary rationality that governs political behavior. Political activity (such as participation in a regime-style coalition) is governed by a whole range of rationalities and irrationalities, which vary systematically according to the kinds of actors and institutions involved. What makes sense to

a business person will be very different from what makes sense to a local politician or community group.

2. Rationalities and the knowledges in which they are based change over time. The building of a regime alters the field in which agents act. In the United Kingdom for example, local voluntary sector organizations have been drawn into increasingly formal and contractual relations with the local state as surrogate service providers. This clearly has the potential to alter the norms and understandings (the common sense) through which such organizations interpret their political world. Table 7.1, therefore, needs to be read as a snapshot of a dynamic process.

3. Practical sense and habitus are stratified and differentiated according to the different fields in which they are effective. In one sense the sphere of urban politics is a single field, with its own particular norms and habitus, but it may be better to see urban politics as constituted at the intersection of a series of different fields (local government, business, community, public and voluntary sectors, and so on) each of which provides a different set of understandings, discourses, and knowledge. This suggests that the problems of regime construction are significantly more complex than that of overcoming a free rider problem among nongovernmental actors. What is involved is no less than mediation, negotiation, and translation between a variety of different practical logics, world views, and ways of knowing.

4. Once a regime is established it may form its own field with its own habitus as actors are drawn into a new set of shared assumptions and practical understandings. Where a regime habitus emerges with a good fit to the regime field (that is, regime practice is well oriented to the regime's conditions of existence), there is scope for an enduring urban political coalition, the explanation for which does not depend on the notion of selective incentives.

5. Fields may be understood as potential "sites of regulation" (or counter-regulation). Processes of regulation operate to promote system reproduction through time. Thus, an enduring fit between habitus, practice, and field that is also contingently effective in stabilizing some aspect of the social whole provides a way of accounting for the emergence and development of regulatory processes without resorting either to functionalism or to voluntarism. This formulation can therefore help resolve the explanatory problems posed by the limits of regulation theory that I outlined at the beginning of the paper.

6. In Bourdieu's work, the notion of field is a *social* concept designating a part of the social whole within which habitus is effective. Fields also have a *spatial* structure relating to both their scale and scope, however. The field of the urban regime operates at the urban scale and incorporates within its scope a limited range of the agents in the city. The fields from which the participants

come, however, can be very different in scale and scope. A manager in a global corporation operates in a field that is very different (socially and spatially) from a community political activist. I previously suggested that the urban scale was not an inherent difficulty for the regulation approach. If fields are understood as potential sites of regulation, then the fact of their differing spatialities holds out the possibility of a spatially sophisticated regulation theoretic account of urban politics, in which a reworked regime theory can play an important explanatory role.

## Conclusion: Strategy, Rationality, and Urban Politics

In conclusion, I want to address a potential criticism of the application of Bourdieu's ideas to *political* practice. The notion of habitus may help understand how the practices of everyday life are related to social structures, but it might be argued that political practices are unlike the practices of everyday life in being the *product of strategic calculation* and are therefore more amenable to the forms of analysis based on the rational actor model that Stone implicitly adopts and Bourdieu explicitly rejects.[10]

*Strategy* is a term that is probably undertheorized in the literature of social and political theory. In a debate in the pages of the journal *Sociology* (Crow, 1989, pp. 1-24; Knights & Morgan, 1990, pp. 475-483; Watson, 1990, pp. 485-498), Crow (1989) argued that using the term *strategy* usually involves adopting (at least implicitly) a rational choice perspective because it carries connotations of conscious decision making and the pursuit of rational objectives. Other protagonists in the debate, however, adopted other positions. William Watson for example argued that strategy need not be understood as involving rational calculation and that some strategies are value-figurative, rather than purposive-rational (1990). Knights and Morgan (1990) argued in favor of making strategy an object of social analysis, rather than a tool thereof. This allows us to distinguish two uses of the concept of strategy. First, it can be understood as a phenomenon in the social world that can be analyzed through the Bourdieusian notion of practice. Second, it can be understood as an analytic tool, and here Bourdieu has his own notion of strategy that he carefully distinguishes from the rational choice model. I will briefly consider each of these.

Although I do not in the end agree with Knights and Morgan's rejection of the concept as an analytical device, their argument provides an interesting link with my previous discussion that returns us neatly to the problems of Stone's account of selective incentives. Knights and Morgan argue that as an object of social analysis, the concept of strategy cannot be seen as a generic category but must be understood in relation to those areas of social life in which strategic action or "strategizing" is a part of social practice. They identify military conflict and business organization as two areas in which the concept of strategy is used self-consciously and where it forms a part of the content of practice, rather than being a conceptual framework within which to understand practice. Although they do not focus on urban politics and political organizations, clearly the concept of strategy is also used substantively in developing political practice.

As substantive phenomena political strategies *do* involve a rational choice style notion of means-ends rationality. Indeed the arguments of rational choice theory can be drawn on *by actors themselves* in developing strategies and in understanding what they are doing as strategic. Does this then mean that such behavior can be explained by rational choice approaches? From the Bourdieusian perspective, the answer is still no because the practice of strategizing is in principle a practice like other practices: generated through habitus and enabled by dispositions that derive from structured and structuring structures. According to Bourdieu,

> It is, of course, never ruled out that the responses of the "habitus" may be accompanied by a strategic calculation tending to perform in a conscious mode the operation that the habitus performs quite differently, namely an estimation of chances presupposing transformation of the past effect into expected objective. But these responses are first defined, without any calculation, in relation to objective potentialities, immediately inscribed in the present, things to do or not to do, things to say or not to say, in relation to a probable, "upcoming" future which . . . puts itself forward with an urgency and a claim to existence that excludes all deliberation. (Bourdieu, 1990, p. 53)

If the approach adopted by Bourdieu is applied to urban politics, therefore, the question changes from

Do agents act in their own rational self-interest (through selective incentives)?

to

How is that particular form of political agency and political subjectivity generated that seeks to calculate its rational self-interest and aims to act strategically to enhance it?

The concept of habitus can be used to examine the production of different types of political subjects (some political subjects engage in strategic practices, whereas others do not). It marks rationality and responsiveness to selective incentives as phenomena to be explained, rather than the source of explanation.

Bourdieu's ideas of habitus and practice imply that *strategizing* is a particular type of practice generated among certain social groups (e.g., politicians, business leaders) by a particular habitus, but Bourdieu also has a concept of *strategy* that he explicitly differentiates from the rational choice model. In an interview he argued that

> far from being posited as such in an explicit, conscious project, the strategies suggested by habitus as a "feel for the game" aim . . . toward the "objective potentialities" immediately given in the immediate present. And one may wonder, as you do, whether we should then talk of "strategy" at all. It is true that the word is strongly associated with the intellectualist and subjectivist tradition which has dominated modern Western philosophy, and which is now again on the upswing with R[ational] A[ctor] T[heory], a theory so well suited to satisfy the spiritualist *point d'honneur* of intellectuals. This is not a reason, however, not to use it with a totally different theoretical intention, to designate the objectively oriented lines of action which social agents continually construct in and through practice. (Bourdieu & Wacquant, 1992, pp. 128-129)

Strategies in this sense are paths across fields, the directions of which are determined by habitus but that can result in changes in the nature of the field. Where two or more fields intersect, as I have suggested is the case in the formation of an urban regime, there is scope for the habitus to be disrupted and for other strategies to be developed in relation to other fields. Agents bring to the regime the cultural assumptions designated by the habitus of their own field, but the disjuncture between these and the other fields involved provides a potential source of dynamism and (strategic) political change.

In this chapter I have not suggested that the work of Bourdieu can provide an easy solution to the problem of reconciling urban regime theory with regulation theory. I do think, however, it can provide a starting point for reworking regime theory in ways that remove its rationalist assumptions, which are some of the main stumbling blocks to an effective dialogue between the two. Although the impacts of urban regimes on local economies can be assimilated effectively within a geographically sensitive regulation approach, the conventional account of regime formation is more problematic. The concept of habitus problematizes the explanatory variables in Stone's account of regime formation and thus opens the door to a version of regime theory that gives full weight to a whole range of different forms of potential political practice, based in a diverse set of political knowledge, understandings, rationalities, and subjectivities.

## Notes

1. I use the terms *counterregulation* and *counterregulatory* to refer to social relations, processes, and practices that tend to undermine or disrupt regulation. Because regulation is not automatic or guaranteed in a structural-functionalist manner, social systems involve processes that tend to generate crises of integration and regulation as well as those that tend to generate system integration and regulation. The extent to which any particular system is actually regulated will depend on the mix and interaction between regulatory and counterregulatory tendencies.

2. I share with Bob Jessop (1992a) the view that the term *fordism* is most appropriately used to refer to a mode of regulation (as opposed to, say, a type of labor process or a regime of accumulation).

3. Mark Goodwin and I (Painter & Goodwin, 1995) have argued that contemporary forms of uneven development can undermine the prospects for the emergence of coherent modes of regulation in the future. As I suggested earlier, however, the regulation approach does not stand or fall with the concept of mode regulation.

4. At this point the argument is entirely compatible with the regime approach, the focus of which is so often the impact of urban politics on economic growth and development. As I shall show, however, the overall compatibility of the two approaches is more questionable.

5. Of course, new regulatory processes and modes of regulation do have effects that in turn condition the character of regulation in a process of mutual constitution. What I am concerned with here, however, are the changes that lead to the emergence of new modes of regulation in the first place.

6. I am grateful to Mickey Lauria for clarifying this point and suggesting the phrasing.

7. In addition to the ideas presented in the remainder of this chapter, some other potential reworkings are developed in other contributions to this book.

8. I have not rehearsed the general arguments against the rational choice approach, partly because they are well known, but more important because my principal interest here is in commensurability with regulation theory, rather than a general critique.

9. Alex Demirovic (1988) argues that the concept of habitus is also incompatible with regulation theory. Unfortunately, I do not have the space to address his arguments here, but in any case, I think that the idea *is* worth pursuing because of the questions it raises about mainstream accounts of political practice. It may well be that the concept of habitus, like that of regime, also requires adaptation. I am very grateful to Bob Jessop for providing me with a copy of his own translation of Demirovic's unpublished paper.

10. Notwithstanding his strenuous denunciations of the rational actor model, Bourdieu's own formulation is seen by some as still caught at least partly within the rational choice framework (Calhoun et al., 1993).

# PART 3

# Concrete Research

## Regulating Urban Politics
## in a Global Economy

Although the authors in this section do not specifically use the theoretical developments discussed in the previous chapters (an inherent complexity of edited volumes), they do construct their analytic frames with what they view as the complementary insights of regulationist and urban regime theories pushed to the forefront. Horan argues that urban regime theory needs to and that regulation theory is useful for refocusing us on the problem of politically bridging the gap between market and state in urban governance. She argues that although the two theoretical approaches have a common focus—the contingent nature of urban governance—the regulationist conception of governance involves the more complex task of social-political-economic harmonization than does the urban regime idea of the political coordination. At the same time, Mann's concept of infrastructural power focuses Horan's attention on how local governments can contribute to spatial advantage. Thus, she evaluates the historical development of Boston's governing coalition in its market context, the shifting of the tax burden distribution, the reorganization of the local state administrative and legislative apparats and their relationship to citizen participation and electoral coalitions.

Beauregard interprets the development of Philadelphia's postwar urban regime within the U.S. postwar regime of accumulation and mode of social regulation. He argues that Philadelphia's regime was not solely the result of local motivations, capacities, and cooperation but, rather, these capacities were facilitated by a shift in the regulation mode within capitalism. Subsequently, when that mode of regulation began to break down, Philadelphia's development-oriented regime began to unravel. Although the objective conditions that demanded reinvestment continued to worsen, publicly led reinvestment diminished. Beauregard argues that the postwar labor-capital accord that provided greater governmental involvement in social welfare activities was important in supporting local progrowth governing coalitions. Local government became an object and agent of this mode of social regulation. Its charge was to simultaneously facilitate and ameliorate the negative externalities of uneven spatial development through slum clearance, urban renewal and public housing, and welfare state services. As the fordist regime of accumulation began to falter and industrial restructuring weakened this labor-capital accord, its associated mode of social regulation became counter productive. The federal government began its withdrawal from welfare state activities, and local governments were forced to become more entrepreneurial to hold their position in the urban governing coalition. At the same time, global industrial restructuring began to weaken the ties of some capital to particular localities. Thus, the governing coalitions began to reorganize, and regimes began to break down.

Keating challenges the claim that Cleveland is the "comeback" city—that Cleveland reversed its decades of decline and is now revitalized. Keating focuses on intercity competition for professional sports fran-

chises as only one aspect of cities being the last entrepreneurs. He argues
that his analysis of Cleveland's governing coalition's strategy for downtown redevelopment epitomizes the insights provided by the urban
regime approach: the primacy of public financing (local, state, and
national) of large private development projects, the role of large economic interests in determining the growth agenda and form of specific
projects, the ideological and consensus-seeking role of local politicians
in selling the projects to the public, the creation of quasi-public entities
with minimal public accountability and oversight to plan and implement
the development projects, and the lack of national reglementory intervention in local place-making competitions in the United States. Thus,
the Cleveland case epitomizes the regulation of urban politics and urban
development projects in a global economy where private capitals reduce
economic risk, increase potential profits, and control the development
arena, and thus their competition, through urban politics. Although
Keating indicates that Cleveland is not unusual here, he also points to
the contingent nature of this urban politics by referring to cases (San
Francisco and Seattle) where the urban regime's electoral coalition was
undermined via local initiative.

   Finally, Jonas argues that the regulationist and urban regime approaches are compatible. Although regulationist accounts of urban politics make connections between transformations in the social regulation
mode and transformations in the form and content of local government,
urban regime theory provides a richer picture of the local interests,
struggles, and strategies involved in urban political transformations. This
position is reminiscent of Painter's theoretical discussions that argue for
the need to focus on how concrete social practices constitute regulatory
processes. Jonas also focuses on the role of territorial scale (spatiality)
in the regulation of urban governance. As were Feldman, Jessop, Goodwin and Painter, and Painter, Jonas is critical of urban regime and
regulation theory's ambivalence toward scale and its causal properties.
His analysis of the reorganization of local governance, particularly in
relation to land development and redevelopment, demonstrates the constitutive causal power of geographical scale. Spatial scale played a role
not only in political coalition building but also in state institution

reformation and in the political coordination of economic development strategies. The urban politics, emerging regime, in Southern California that Jonas describes may be one of the emerging political economic forms in the postfordist regime of accumulation.

# 8

# Coalition, Market, and State

## Postwar Development
## Politics in Boston

CYNTHIA HORAN

In the past 15 years, development policy has become a central focus for urban political analysis and regime theory has become the dominant framework for investigating urban development policy making. Starting with the claim that "local politics matters," regime theory has informed numerous studies of urban development since World War II and inspired a body of empirical scholarship essential for understanding what has been, and is, happening to urban political economies.

This chapter, too, is informed by regime theory—more precisely, this chapter results from a debate with regime theory. A product of investigating urban development politics and economic change in Boston since 1945, the chapter will argue that although the urban regime paradigm illuminates key aspects of development politics, this paradigm remains too narrow a framework for investigating the processes of urban political and economic restructuring. To reconsider regime theory, I will reflect on another, quite different, approach to economic restructuring and politics—regulation theory. Comparing the analysis of "local governance" in both theories offers a way to both critique and broaden the research agenda of regime scholarship.

I will argue that regime scholarship would benefit from a reemphasis on one of its early propositions: that city politics are structured by the division of labor between market and state (Elkin, 1985, 1987). Governing coalition politics, business politics, electoral politics, and oppositional politics are both framed by, and frame, the interrelated dynamic of market and state. On the basis of a critical reading of regime scholarship, the Boston case demonstrates why and how analyzing market-state relations matters to understanding both economic restructuring and urban politics.

## Governance and Urban Political Economy

### Urban Regimes and the Analysis of Governance

It is by now unnecessary to point out how urban regime scholarship began with the message that "local politics matters." Although its many practitioners and case studies make generalizations about regime scholarship difficult, most of us would probably agree with Clarence Stone that

> regime theory . . . has two basic premises: (1) development policy is not determined in all its particulars by the mobility of capital, and (2) that, within the constraints of a profit-based economy, local development policy is shaped by the nature of the governing coalition. (Stone, 1991, p. 290)

Several characteristics of regime theory are apparent here. First, regime scholarship is particularly concerned with development policy choices. Second, it understands those choices within a more or less determining structure of an increasingly globalized capitalism as well as limited legal and fiscal powers, metropolitan political fragmentation, and own source revenue reliance (Elkin, 1985, 1987). Third, local governing coalitions of diverse interests, not "unitary" actors or global "forces," effect development policy. Crucial to understanding governance in regime theory is analyzing the relationships between structures and choices, and crucial to that task is the examination of urban governing coalitions.

Regime theory could be portrayed as the study of urban governing coalitions though its understanding of governing coalitions has, I think, changed significantly over time. Initially, regime scholarship argued quite explicitly that a dual, potentially antagonistic, logic of power characterized American cities: control of economic assets and electoral success—what Stephen Elkin called the division of labor between market and state (Elkin, 1985). The governing coalition both literally personified that dual basis of power and harnessed it by overcoming, through cooperation and negotiation among economic and political elites, the division of labor between market and state.

The urban governing coalition was itself a complex product of structure and choice as its uneasy, often undemocratic relationship to its electoral constituencies made clear. In Stone's words, a central question for regime scholars is "Why do most cities embrace the development agenda of downtown business even though that agenda neglects and even harms the interest of the nonbusiness majority?" (Stone, 1991, p. 290). Critical of pluralism, the bulk of early regime scholarship demonstrated how governing coalitions controlled development policy despite the threat of electoral defeat and community struggles. Development policy making was political because it encompassed choices to preserve the governing coalition and to prevent non-elite protests.

Though a critique of pluralism, regime theory was never clear, I think, about whether or not capitalism, that is, the market, necessarily advantaged business interests in urban regimes whatever their participation in governing coalitions. In that sense, the relationship between market and state remained ambivalent or perhaps under-theorized. On the one hand, most regime case studies document the regressive consequences of most development projects. On the other hand, many regime scholars seem to believe that, as Stone put it, "the economic imperative is pliable within limits. . . . Within the bounds of that imperative the terms of cooperation between government and business are negotiable" (Stone, 1991, p. 290).

Over time, the structural bases and logics of governing coalitions have attracted less and less attention. Instead, scholars now focus on the negotiations that keep governing coalitions intact or make new coalitions possible. Further, though always a concern, an increasingly important theme in regime scholarship is the difficulty of urban governance itself. The governing coalition takes on a new purpose: it solves collective action problems and makes governing possible. No longer an elite power relationship in which business is privileged, the urban governing coalition has become a valued and potentially more inclusive form of rule.[1]

In and of itself, this concern with governance seems a logical corollary of regime theory's focus on policy choice. At the same time, it seems to me that in its recent treatment of governance, regime theory is straying from one of its most intriguing and original propositions, one that not only distinguished it from other urban politics paradigms but also made it a potentially rich framework for investigating economic restructuring: that city politics is a product of the historic, dynamic, and specific division of labor between market and state. In that earlier formulation, urban regime analysis necessarily involved attention to the politics of structures as well as to the politics of coalitions or interests.

Further, I would argue that a focus on governance without a concern for the structures of market and state simply cannot work. Consider, for example, how the market might shape local policy choices. The specific economic context facing the governing coalition presumably shapes the possibilities and consequences of its development programs and thus should be carefully investigated. In addition, because regime theory emphasizes that the mobility of capital is a potential, though contingent, constraint on local politics, analyzing the ties of businesses to specific cities appears to be a crucial step for regime analysis.

And, if the Boston case is any guide, we cannot focus our attention only on the market side of the division of labor. Here Elkin's discussion of local political institutions is helpful. Although emphasizing the legal and fiscal limits of cities as states, Elkin nevertheless argues that local political institutions matter to governance: they create public bureaucratic and regulatory capacities, define the parameters of representation and negotiation, and constitute political preferences and interests, both public and private. Political institutions thus help create and empower urban regimes; they define the context and substance of urban governance, rearranging them is a principal focus of regime politics.

## Regulation Theory and the Analysis of Governance

Regulation theory, too, examines local governance. Although its concern is more recent and thus we cannot draw on a large empirical scholarship, we can nevertheless see certain similarities between regulation and regime theories. For regulationists, as for regime scholars, the study of local governance involves analyzing the exercise of authority by more or less independent governmental and nongovernmental actors and organizations. Both approaches thus ask a broadly similar question: How (around what agendas and projects and on what terms) is a diversity of actors coordinated so that collective issues, particularly the problem of economic development, can be addressed? Both emphasize the contingent nature of governance outcomes and thus the process, or practice, of governing is a focus. Both theories situate local governance in the dynamics of economic restructuring and fiscal constraints. Despite these important similarities, I find the differences between the theories more enlightening, and I will explore some of these differences here.

In regulation theory, the analysis of local governance is inextricably linked to economic structures and dynamics. More specifically, it is premised on the logics of accumulation, competition, and crisis that characterize and constitute the capitalist mode of production. Studies of local governance are part of a literature that examines the collapse of the dominant 20th century capitalist accumulation regime, fordism, and the emergence of postfordist accumulation regimes. For me, this contextualization of local politics has the virtue of directly engaging the question: "Does local politics matter?"

For regulationists, local politics potentially matters because globalization does not exhaust the accumulation regime of postfordism. Rather, new relations between global, national, and subnational economies are emerging—hence the term *glocalization* to denote simultaneous processes of globalization and localization. In its consideration of subnational accumulation regimes, this scholarship is partly rooted in the rich literature on uneven development. Echoing long-standing debates in radical geography, the analysis of accumulation regimes involves considering their propulsive sectors. Different sectors exhibit different location preferences; depending on their economic bases, cities enjoy variable spatial advantage and potentially divergent politics (Tickell & Peck, 1992).

In its debate over whether, and how, relative stable regimes of accumulation can be created and sustained at subnational scales, regulation scholarship moves beyond the proposition that local politics derives from the local dependence of capital (Cox & Mair, 1988). A second important argument in regulation theory is that regimes of accumulation, indeed capitalism itself, cannot be considered as purely economic. As Bob Jessop explained, "[regulation theory] is particularly concerned with the changing forms and mechanisms (institutions, networks, procedures, modes of calculation, and norms) in and through which the expanded reproduction of capital as a social relation is secured" (Jessop, 1990a, pp. 154-155). Regimes of accumulation encompass modes of social regulation, the "institutional complementarities that secure the regularization of accumulation regimes" (Jessop, 1995b, p. 319). Since the institutional complementarities underpinning accumulation regimes are not determined by accumulation, the regulationist conceptualization of uneven development extends to local differences in culture, society, and politics potentially crucial to modes of social regulation and hence to accumulation.

From this perspective, governance can contribute to local modes of regulation and, thus, local governance can enhance spatial advantage (the local accumulation regime). This idea of governance involves a more complex task of social-political-economic harmonization than the idea of political coordination central to governance in regime theory. Furthermore, in contrast with regime theory, governance here involves not only the collaboration of independent interests with differing agendas and variable resources but also the correspondence of structures with differing logics—capitalist, democratic, bureaucratic, cultural, racial, and so on.

Of particular interest to this chapter is that although a focus on governance signifies that government's role is shrinking, for regulationists the state remains central to governance: "Local governance is produced in and through institutions, including, but not limited to, decentralized state institutions" (Painter & Goodwin, 1995, p. 342). Unfortunately, just how governance is produced in and through local state institutions is not well understood. As Jessop recently noted, "there are as yet no adequate *explanations* of the structural transformation and/or strategic reorientation of the local state" (Jessop, 1995b, p. 321). Although acknowledging the importance of local state institutions, regulation scholars recognize they have yet to develop a theory of the local state (Painter & Goodwin, 1995).

## From Theory to the Boston Case

The remainder of this chapter will examine governance in Boston from 1945 to 1992 within the changing structures of the local economy and of the local state. The empirical analysis to follow has been shaped by the multiple perspectives of both regime and regulation approaches. The examination of Boston's governing coalition lies firmly within the regime paradigm. My attention to the economy has been influenced by the literature on uneven development central to regulation theory.

The examination of the local state depends partly on other arguments. I take seriously, and indeed corroborate, Elkin's proposition that because local political institutions can shape political interests, institutional politics is central to regime politics. But I owe a bigger debt to the work of Michael Mann (1984, 1993). Mann's insistence that an examination of state capacity be divorced from a discussion of state autonomy has offered me a way to appreciate other relationships between the local state and the economy. Mann's contention that the crucial power of democratic capitalist states is infrastructural, the "capacity to actually penetrate civil society, and to implement logistically political decisions," whether those decisions originate with state elites or with interests outside of the state (Mann, 1984, p. 5), suggests how even local governments hemmed in by statutory constraints and business elites might wield power. In combination with the literature on uneven development, the concept of infrastructural power alerts us to situations where local governments contribute to spatial advantage. Finally, because Mann stresses the centrality of state institutions to infrastructural power, he underscores the importance of institutional analysis to the study of governing. Although this chapter does not rest on, nor offer, a theory of the local state, I do attempt to show in what ways, in one particular city, local political institutions contributed to economic transformation.

## Understanding the "Market": Placing Boston's Governing Coalition in the Economy

An examination of Boston's postwar development politics must begin with the recognition that the city is one of America's postwar urban success stories. What distinguishes Boston's regime politics from those in many other U.S. cities is that after Boston's postwar governing coalition emerged, the city's economy began to grow. Understanding the role of regime politics in the city's economic transformation and the role of the economic transformation in the city's regime politics is crucial. It allows us to consider a crucial question: How does local politics shape spatial advantage?

The extent and nature of the city's postwar economic change can be suggested by briefly comparing it to other cities. In their analysis of U.S. metropolitan economies, Noyelle and Stanback grouped Boston with similar large, older, regionally dominant cities that offer a useful benchmark for this analysis: Baltimore, Philadelphia, and St. Louis (Noyelle & Stanback, 1984).[2] In this group, Boston was the only city to demonstrate a net gain in private sector employment from 1951 to 1990: a 27.5% increase compared with a 9.7% decline in Baltimore, a 21.2% decline in Philadelphia, and a 38.6% decline in St. Louis. Boston's employment rose in two periods: from the late 1950s to 1969 and from the late 1970s to 1989. Job losses characterized the 1970s and the 1990s. Even after the most recent recession, Boston has 16% more private sector jobs than it did in 1951.[3]

In part, the city's job growth reflects a postwar metropolitan economy boosted by the restructuring and changing location of manufacturing in New England. In a marked departure from 19th century patterns of dispersed production sites, postwar manufacturing firms, first in defense and later in high technology, tended to locate near major scientific universities and in larger urban labor markets—a preference that benefited the Boston SMSA (standard metropolitan statistical area) (Estall, 1966). In absolute terms, suburban job growth in the Boston SMSA ranked second after Philadelphia with roughly 60% of the job increase occurring in the high technology expansion of the late 1970s to 1980s.

Yet, since Philadelphia demonstrates huge job losses despite even greater suburban employment expansion, what matters is Boston's ability to compete with its suburbs. That edge primarily lies in the city's attractiveness for producer-services firms.[4] The city began the postwar period with a smaller dependence on manufacturing employment and, more importantly, with a larger share of jobs in the economic sectors that proved crucial to post-industrial urban economic growth: finance and business services. Over the postwar period, as its share of total private sector employment climbed to 61% in 1990, the city's producer-services economy shifted from an historic specialization in finance to a more diversified base. Today, measured in both job share and in the absolute number of jobs, Boston has one of the largest producer services economies in the country. To continue the earlier comparison, although Philadelphia has 20% more jobs than Boston, Boston's producer-services employment is 27% greater (U.S. Bureau of the Census).

As this services economy has expanded, the city has managed to remain significant to the growing metropolitan economy. Starting the postwar era with a highly suburbanized economy,[5] Boston's share of metropolitan employment has to continued to drop; today the city houses about 23% of SMSA employment. But by other indicators, the Boston SMSA remains centered on its major city. First, the city itself remains one of the few central cities to house more jobs than people (Boston Redevelopment Authority [BRA], 1992). It is the site of one third of the SMSA's producer services employment, not much below its 1970 percentage, even as these services grew rapidly in the metropolitan economy particularly in the 1980s. According to the BRA, over the last 15 years, the city's firms and workers have become increasingly productive, outpacing suburban productivity gains. By 1989, with 17% of the state's employment, the city accounted for 24% of its economic output (Perkins, 1990).

Further, the city's downtown office market continues to dominate the SMSA office economy. From 1960 to 1990 the city's supply of prime office space doubled, making Boston the fourth largest office market in the country, up from seventh in the 1950s (BRA, 1992; Schultz & Simmons, 1959). The centrality of the central business district has been enhanced by an extensive, city-focused public transit infrastructure, the major New England airport's downtown Boston location and even by postwar highway building that, though adding circumfer-ential roads around the city, also emphasized radial roads into its core. Despite substantial suburban office construction in the 1980s (the city now houses about 50% of prime SMSA office space), the city's downtown has remained attractive to investors, local and otherwise. As one major investment guide put it, "The continuing suburban sprawl diminishes the benefits of being outside the CBD" (Salamon Brothers, 1989, p. 4). The postwar period demonstrates two major periods of office investment and, although the data are limited, the two periods differ in ways suggesting the changing place of the city in national and international property markets. In the 1960s and early 1970s, the investors were primarily, though not initially or entirely, Boston firms building for themselves and for specific tenants or owners. In the 1980s, speculative investors, often major national or international developers and financial firms, fueled the building boom (BRA, 1976; Salamon Brothers, 1989).

The contribution of Boston's postwar governing coalition—as a political alliance—to the city's economic transformation was the rejuvenation of the city's downtown property market. Such a rejuvenation entailed not merely the selective demolition of a crumbling built environment and a massive infusion of public and private money. It also involved creation of new political relations, both informal and institutional. The most important steps in this political reorganization took place in the city's first period of economic expansion as the city's downtown business elite and its political leaders negotiated a new alliance. Changes instituted in the 1960s, combined with the massive production of new modern office space, itself the result of these political and institutional changes, created a downtown office economy that in turn attracted even more real estate and non-real estate investment in the 1980s.

In retrospect, the 1950s and 1960s decades are an important moment of economic as well as political change. After 50 years of minimal commercial construction, in 1959, downtown office space expansion began that continued, with a slowdown in the 1970s, until the late 1980s. Also in 1959, after falling for 30 years, city per capita income began a sharp upward rise in real terms that has continued ever since (Carlaw, 1979). Finally, at the start of these decades, the city's job base remained decidedly early 20th century. Manufacturing was the largest source of jobs with roughly three times the employment of finance, insurance, and real estate (FIRE); business services accounted for less than one fifth of services employment, about the same as personal services. By 1970, the employment structure was much like it is today. FIRE employment had surpassed manufacturing (medical employment was almost as large as manufacturing); business services provided more jobs than either banking or insurance and almost four times the jobs of personal services (Foley, 1990; U.S. Bureau of the Census).

## Organizing a New Governing Coalition

The postwar political mobilization of Boston business was catalyzed by an economic dilemma made clear in a 1959 report funded by the city's major downtown firms:

While the downtown remains the most functional place in the metropolitan area to which to attract a large clerical labor force, the property valuations and taxes make substantial new investment in space to house such operations downtown almost prohibitive. Yet many of these companies, the utilities, the large banks and life insurance companies, already have considerable investments in fixed facilities which makes it difficult for them to abandon the downtown entirely even if this were operationally feasible. The resulting impasse, in which it is equally impossible for businesses to expand and equally impossible for them to leave, represents a succinct summary of one of the major threats to the future of downtown Boston. (Arthur D. Little, 1958, p. III-10)

We see here a classic statement of locational dependence and a dependence that necessitated political mobilization. Without political action, how would taxes be lowered?

Boston's postwar progrowth governing coalition thus began in a business-initiated drive to oust Mayor James Michael Curley who had dominated Boston politics for decades (Horan, 1990; Marchione, 1976; Mollenkopf, 1983; O'Connor, 1993). At the core of this reform effort were several of the city's leading downtown firms, whose executives embodied not only an economic elite but also a social one: the First National Bank of Boston, National Shawmut Bank, Boston Safe Deposit Company, John Hancock Insurance Company, New England Telephone, Boston Edison, and several department stores. A few prominent attorneys in old Boston law firms played key roles in organizing businessmen and brokering deals with politicians.

For business, Curley's regime was a disaster. Less a political machine than a set of loosely connected ward factions competing for the spoils of office, Curley's spending priorities were a mix of vote buying and redistribution (Beatty, 1992; Erie, 1988). Curley reoriented city planning projects away from the downtown: schools, housing and neighborhood facilities became a major priority (Kennedy, 1992). An army of workers, contractors, and lawyers made Boston's government one of the most costly in the country. Even worse, Curley's view of the economy was essentially predatory; since the downtown was owned by the Brahmins, promoting economic growth would simply enhance their economic power. The mayor's frequent threats against State Street, the financial district, were not rhetorical: Boston had the steepest business property taxes of any big American city.

First evident after World War II, organized business political activity continued throughout the 1950s and 1960s. Although business's dislike of Curley was long-standing, Curley's ebbing support among middle-class voters and the decay of ward political organizations presented an opportunity unparalleled since the early 1900s. In 1946, key downtown businessmen spearheaded an election campaign and charter reforms to end Curley's career. As his successor, John Hynes (1948-1959), demonstrated that a probusiness mayor did not ensure probusiness policies, downtown executives worked hard to mobilize business-men politically through the revitalization of existing organizations (the Chamber of Commerce), through the takeover of nonbusiness groups (the Massachusetts Taxpayers Federation) and through the creation of new business groups (the Greater Boston Economic Study Committee [GBESC] and the Coordinating Committee or Vault, which remains a political organization of major downtown firms). Alliances with the city's Irish establishment were also slowly crafted. By 1959, business had become the most powerful political interest in the city. This mobilization of business made the new regime possible; it framed its goals and, I will argue, the principal institutions for realizing those goals.

The anti-Curley politics of business should not be seen simply as another chapter in the saga of machine versus reform politics in American cities, though, like prior reform movements, businessmen intended to make government more efficient and less corrupt. Working within a regime framework, I believe that, after the war, organized business groups had a more ambitious goal: they sought an altogether different relationship between government and the(ir) local econ-omy. The new relationship involved tax reform. But it also involved a govern-ment that would work for business.

A probusiness governing coalition is most visible during Collins's adminis-tration (1960-1967). Collins met regularly with the Vault and made downtown renewal the centerpiece of his administration (O'Connor, 1993). After Collins's departure, the close working relationships between city hall and business organizations ceased and, I will argue, the governing coalition became more state-centered—dominated by mayors and their top bureaucrats.

Just why business organizational participation declined is a matter of some dispute. A common explanation points to shifting mayoral priorities: Collins's successors Kevin White (1968-1983) and Raymond Flynn (1984-1993) both campaigned against corporate center development strategies and, by implica-tion, against business. But since both mayors worked closely with businessmen and organizations on other issues, notably education, they clearly found business support helpful and politically acceptable. Offering another perspective, *The Boston Globe* has frequently speculated that the takeover of Boston firms in the regional boom of the 1980s resulted in a less Boston-oriented and less politically active group of executives (Snyder, 1988). Business mobilization on develop-ment issues declined well before the 1980s, however. Without completely discounting the effects of either mayoral agendas or corporate takeovers, I will add another explanation: the early political and economic successes of the postwar governing coalition made sustained mobilization less important (see also Ehrlich, 1987). In the 1960s, business priorities became the priorities of government; whatever the results of electoral politics, key institutional reforms reinforced the agenda of business. We must thus consider how the reorganization of the state sustained the transformation of the economy.

# Reorganizing the State: Shifting the Tax Burden

The postwar political alliance between businessmen and mayors was initially secured through, and embodied in, a new tax structure that made the city's high property taxes competitive with those of other cities. As the earlier quotation from the GBESC study suggests, slashing property taxes was the major reason for business's political mobilization. Unable to secure property tax relief through state legislative or judicial action in the 1940s and 1950s, business finally obtained tax reform after a bitter struggle between business organizations and city councilors representing middle-class voters (Horan, 1990).

This new tax policy epitomized the informal practices that regime theory sees as crucial to governing. It represented a political compromise among the interests crucial to the coalition: downtown businesses hoping to expand their operations and more affluent homeowners. Business's preferred tax reform, reassessment of all property at market values, would have shifted burdens onto affluent homeowners and thus was defeated. Instead, a dual policy of tax concessions for new construction was adopted.

The reform was itself the product of delicate negotiations between business organizations, city officials, and state government politicians and bureaucrats. The legislated form of concessions was restricted by state law to publicly regulated projects in blighted areas. The other, more commonly granted, concession was implemented through private letters between mayors and developers of large commercial projects (Horan, 1990). Unregulated by law (hence, private), these concessions were illegal because—in violation of the state constitution—the city was never authorized by the state legislature to grant such concessions (Ragonetti, 1977).

Given their questionable legal status, the informal concessions' continued survival depended on a set of informal practices to prevent court challenges. Mayors agreed that concessions would benefit all new commercial construction; business groups worked to prevent lawsuits by disgruntled owners of existing commercial properties that continued to pay high taxes; both political and economic elites managed to hide the policy from the public, whose residential tax bills rose as downtown tax bills fell (Horan, 1990; Ragonetti, 1977).

In addition to their political importance, tax concessions had major economic effects. First, the benefits of concessions were substantial; they cut downtown office building tax burdens almost in half (Slavet, 1977). That drop proved crucial to the first period of downtown investment. No new office building was constructed until tax concessions were in place and, as far as I can determine, every office building constructed between 1958 and 1982 received a tax concession (Ragonetti, 1977; Slavet, 1977). Favoring new office buildings and big firms in the downtown, these concessions shifted tax burdens onto other types of uses and other types of property owners: small businesses in older buildings, manufacturing firms, and renters. Finally, tax concessions resulted in revenue losses for a city continuing to suffer fiscal problems. As one HUD-financed study concluded in 1977, "the City has not been able to capitalize fully on the extraordinary level of downtown growth" (Slavet, 1977). By Flynn's election, a series of court decisions and the passage of a statewide property tax limitation law (funded by high-tech firms but opposed by Boston business organizations) had made informal tax agreements legally more risky, and they were stopped.

Tax reform thus cemented the informal alliance between city hall and business and, given the city's attractiveness for services firms, also triggered substantial new construction. Though, as we shall see, Boston's businessmen endorsed urban renewal, their vision of economic growth always rested on private, not public, investment. However politicized and legally questionable, from business's perspective, the new tax structure represented a substantial improvement in business-government relations.

## Reorganizing the State: Building a Bureaucracy

As late as 1958, Boston's government had little, if any, capacity to revitalize its property markets. The city's urban renewal program was a disaster; its one major project, the West End, had prompted public outrage over the forced exodus of 10,000 people coupled with charges that Mayor Hynes had improperly awarded the cleared land to a former aide. Alert to redevelopment successes in other cities, many of Boston's traditionally conservative businessmen had accepted the need for a "flexible" development authority "unhampered by bureaucratic red tape and legislative restraints" and responsive to the market (Arthur D. Little, 1958). Hiring one of the country's preeminent public developers, Edward Logue, the postwar governing coalition managed to create a development bureaucracy, the BRA, which surpassed the hopes of its sponsors—as their reminiscences make clear (Hodgkinson, 1972; O'Connor, 1993). By the end of the urban renewal program, Boston ranked first in urban renewal funds per capita and fourth in total funds received (Real Estate Research Corporation, 1974).

The importance of the BRA lies not only in its massive urban renewal program, although the priorities and scale of that program sustained and subsidized private sector office investment in the 1960s. Also significant is the agency's capacity to implement the goals of the 1960s through economic downturns, massive federal spending cuts, and community challenges. The BRA formulated, implemented, and institutionalized the corporate center strategy. Like the new tax structure, the agency both embodied and enabled the new relationship between city hall and business.

Since Logue, the BRA has been a highly effective public agency in a city not known for them. The city's prosperity—for which the BRA did, and can, claim credit—has helped enormously. Its mission tied to fostering the city services economy, the BRA's status was enhanced by economic growth and, more recently, diminished by economic stagnation. Following Mann (1984), I would argue that the BRA worked to enlarge its infrastuctural powers over the city's property markets. Benefiting from a series of strong directors and a well-respected professional staff, the agency has made good use of its regulatory powers over land and buildings, its near monopoly of public development expertise, and its ownership of property acquired with urban renewal monies, all of which were made more valuable by the increasing attractiveness of the city's financial district. For these reasons, much as regime theory would predict, the BRA became a valuable ally for mayors, even those who had won office campaigning against it.

It would be correct to see the BRA as an instrument of mayoral power—but only partially correct. In Boston's strong mayor system, the BRA is a mayoral agency. The mayor appoints the majority of the BRA's board and thus hires the BRA director. Exempt from civil service, the BRA staff offers a potential source of mayoral patronage, and its operating budget was, until recently, part of the city budget. Despite this statutory relationship, only one of Boston's mayors, Kevin White, intervened aggressively in agency affairs. Among other actions, White once forced out a BRA director who had refused to award a BRA-controlled site to one of White's campaign contributors.

Mayoral influence has been mediated by the BRA's mission as the city's quintessential public-private partnership; whatever the mayoral agenda, BRA directors have consistently defined their job as promoting downtown property values and the corporate services economy, a vision reinforced by the local press and business organizations. The agency's downtown priorities date from the heyday of business-city hall collaboration in the 1960s. To draw up Boston's downtown urban renewal plans, Logue's BRA staff worked with two Chamber of Commerce committees, sharing the costs of consultants and agreeing to a business veto over urban renewal applications. From the start, a clear division of labor was established between public and private redevelopment in the downtown. In business's domain, the financial district, government intervention would be minimal although major public funds would be used to redevelop adjacent "blighted" areas. BRA planning clearly identified the city's economic interests with those of its downtown corporate services firms although manufacturing was still the city's largest employer.

Despite White's initial campaign assault on urban renewal, the agency's commitment to private investor agendas did not end with Logue's departure. Although the press has depicted White's BRA as an embattled, enfeebled bureaucracy bending to the mayor's will (demonstrated by the resignation of the BRA director mentioned earlier) (King, 1990), more important for this study was the BRA's ability to pursue the 1960s plan without its funding assumptions. In the 1970s, a downtown vision reliant on public money was transformed into a market-driven development strategy despite predictions that such a strategy would never work (Real Estate Research Corporation, 1974). As federal funding dried up and office development slowed, the BRA shifted from public builder to public deal maker. It successfully implemented some of the country's largest redevelopment projects, notably Quincy Market and the Charlestown Naval Yard (thus not only shoring up downtown property markets but also extending the area of profitable investment outside the downtown's traditional boundaries) (Brown, 1986).

Although fostering private investment, the BRA's planning priorities encompassed more than the immediate concerns of business. Logue, for example, saw the city's universities and hospitals as future sources of economic growth and facilitated their expansion; business organizations opposed such tax-exempt building as driving up property taxes. Logue stressed the importance of rehabilitation and historic preservation, the need for public spaces particularly in the waterfront, and the significance of residential as well as commercial uses to a viable downtown. Although these policies eventually enhanced the city's economic attractiveness—as investors now acknowledge (Salamon Brothers, 1989)—because these policies restricted private decisions, they were not popular with business.

The agency's continued capacity to integrate a public agenda with market objectives is demonstrated in its design and implementation of Flynn's balanced development policy in the 1980s. Like White before him, Flynn campaigned on a pro-neighborhood platform. Unlike White, Flynn had little interest in the details of development policy which he handed to his BRA chief, Stephen Coyle. Coyle's appointment was seen as a signal to businessmen that the mayor was not hostile to them (Powers, 1986). Indeed, Coyle articulated views quite distinct from Flynn's. Although the mayor emphasized the responsibilities of government, Coyle stressed the virtues of the market. As one study noted,

> The BRA's new approach to development has increasingly resembled that of private development entrepreneurs . . . The BRA advanced its own institutional interests by using the substantial income from groundleases to expand its own operations, hiring new staff and raising agency salaries. (Ehrlich, 1987, p. 101)

So lucrative were these incomes that BRA gained financial independence from the city budget process.

The eventual balanced development policy joined populist procedures with enhanced regulation to use Boston's booming real estate market for social objectives while sustaining that market. Central to Flynn's program was linkage, a policy adopted by White, but expanded beyond its initial focus on funding housing construction to funding job training and development projects by African American and Latino developers in poor neighborhoods. A residents jobs policy was also enacted. Neighborhood committees were given advisory roles in planning and zoning decisions in selected areas of the city. Because most of these policies had emerged from more than two decades of community struggles, they represented political victories for opponents of the BRA's market-driven policies.

Yet the BRA's balanced development policy remained crucially dependent on the very development processes that community groups had long challenged. Linkage fees were paid only by developers of large office buildings. Negotiating with developers for community benefits works only for communities attractive to private investors—not for those most needing jobs or improvement. Moreover, as formulated by Coyle, balanced development rarely encompassed community development in poorer areas of the city even though the BRA acknowledged that "Development outside of the downtown area has become increasingly important as the weak link between investment in the downtown and income in the neighborhoods becomes more evident" (Brown, 1986, p. 39). The BRA's major neighborhood economic initiatives relied on enticing corporate services out of the downtown rather than on fostering the multisectoral, small and community-oriented businesses favored by local groups and demonstrated to hire more city residents at better wages.

My point is not that Flynn ignored community development but, rather, that his principal planning and development agency had a distinct view of economic growth that emphasized "development" as much as "balance" and thus assigned distinct economic roles to the downtown and to the neighborhoods.[6] Not until after the downtown real estate market collapsed in the late 1980s did the BRA dramatically change its focus. Then it began to articulate a policy of promoting manufacturing firms.

One aspect of balanced development has drawn less attention. Even as he exploited the market, Coyle decided to regulate the supply of new construction and to impose drastic height limits in the financial district. Without the public struggles so crucial in San Francisco, Boston became one of the toughest cities in which to put up an office building. Slowing the building process allowed the BRA to use its design review and zoning authority more effectively. Further, by preventing an oversupply of office space, the BRA kept vacancy rates from rising and property values from falling as much as they did in many other cities or, for that matter, in the Boston suburbs (Salamon Brothers, 1989). Though in the end the BRA could not prevent the collapse of the office market, it softened the blows.

Created by the 1960s governing coalition, the BRA thus evolved from a working partner of business into an agency capable of facilitating private investment and of pursuing agendas that, though not at odds with business, did not originate with, nor simply reflect, business goals. A technical agency, the BRA quite clearly defined itself as a political player: In both roles its goal was sustaining the city's downtown property market and hence even more than the tax structure, the agency enabled and embodied the new postwar relationship between government and the economy.

### Reorganizing the State: Managing Participation

The last piece of the governing coalition's political strategy tackled neighborhood politics. Postwar Boston discloses many active community groups and some outstanding examples of successful struggles against the city's progrowth policies. But such activism must be examined within an institutional context: Boston's charter provides limited formal authority to its city council and tough initiative and referenda rules. Decision-making power is highly centralized in the mayor's office. Further, Boston's postwar governing coalition, most especially its political elites, has consistently tried to manage participation. Both institutional arrangements and elite political strategies have worked against the representation of the city's poor communities.

Neighborhood organizing has taken place in an environment of drastic socioeconomic change. Since 1950, more than 200,000 people have left the city, and the poorest wards have suffered the biggest losses. The long decline of manufacturing and the boom in city services employment have had mixed implications for city residents. Postwar growth reversed 30 years of income decline; in the 1980s, city incomes grew faster than in the state and nation and poverty rates dropped to levels two thirds that of all central city rates (Sege, 1992). Since the late 1970s, however, studies have documented how city, as opposed to suburban, residents hold the worst paying jobs in the highly differentiated services economy (Boston Foundation, 1989; BRA, 1978). Although communities of color have benefited least from economic restructuring, the city's less educated whites have not done well either. In October 1994, *U.S.News and World Report,* using 1990 census and Urban Institute data, described South Boston as one of the "worst white underclass neighborhoods" in the country (Whitman & Friedman, 1994).

Boston's postwar governing coalition worked hard to prevent the concerns of the city's poorer neighborhoods—Curley's electoral base—from dominating development policy. Boston's business elites and their middle-class allies saw these communities as a threat to their political agenda as well as to their property values. Their solution, achieved in 1949, was to replace the 22-member district city council, the basis of the city's ward political organizations, by an at-large body. Reformers calculated that citywide elections would alter representation and policy making. Successful candidates would have to be wealthy or depend on the financial support of business (Marchione, 1976).

Although facilitating the governing coalition's agenda at first, the council's limited representation made it unable to respond to the consequences of economic change. In the 1950s and early 1960s, the council overrepresented the affluent wards; after that, a few Irish neighborhoods dominated. In either case, the poorer neighborhoods targeted by city hall development strategies often had no way to voice their demands or to protect their interests. As protests were organized in the streets, the city's political elite sought other mechanisms of working with these communities.

Initially, the BRA was given the job of community outreach. Intending to mobilize support for his policies, Logue made participatory planning a centerpiece of his neighborhood improvement strategy. Because the BRA's dominant objective was to retain the city's dwindling middle class (Keyes, 1969) and, of course, to secure property values through public-private gentrification, community outreach in poor neighborhoods proved difficult.

A widespread, at times spontaneous, but often tightly organized, grassroots opposition arose to fight the BRA's neighborhood agenda during the Collins/

Logue years. Although battling the BRA largely independently, community groups voiced a message that would echo again and again in the subsequent decades: greater representation in decision making and protection of existing communities. Then, as later, community groups resisted the depiction of their neighborhoods as slums and their city as a growth machine. The procedures designed to build support for the BRA thus became a means for community groups to fight, delay, and, in some cases, derail its program (Keyes, 1969; Mollenkopf, 1983). The hostility that flourished eventually drove Collins out of politics and Logue out of Boston. It also helped elect Kevin White and later Raymond Flynn, both of whom campaigned advocating neighborhood causes.

White put the mayor's office firmly in control of neighborhood policy. Fourteen little city halls were set up to serve as ombudsman, community advocate, and communication link between city hall and the restive streets. BRA planners were assigned to them. At first, this structure of neighborhood bureaucracies facilitated the causes of community organizers; Boston's little city halls became a national model of government decentralization. White's staff worked more closely with a wider variety of neighborhood groups than had Logue's BRA (Mollenkopf, 1983). Summing up these years, Mollenkopf wrote, "Neighborhood confrontation with City Hall slowly and painfully evolved into collaboration between 1968 and 1975" (Mollenkopf, 1983, p. 191).

For White, as for Logue before him, neighborhood mobilization soon revealed its disadvantages as a governing strategy. As the escalation of conflict over school desegregation pitted white neighborhoods against black ones, "community control" was redefined. Decentralization became a way "in which issues [could] . . . be handled without causing citywide repercussions" (Weinberg, 1981, p. 95). Since his margins of victory were never large, the mayor was constantly in search of a stable electoral base, one not assured by the militance of African American neighborhoods that had been key to his first victory. Slowly, the little city halls became a way to co-opt and manage community leaders and organizations; White increasingly targeted patronage to middle-class areas that, as under Collins, became the electoral basis of mayoral power (Jennings, 1986; Katz, 1978). A changing set of mayoral bureaucracies funded by federal dollars was created to oversee various community development programs throughout the 1970s (Katz, 1978; Yin, 1980). Increasingly enmeshed in bureaucratic procedures, community groups saw their influence whittled away during White's tenure.

Yet the city's grassroots politics also resisted incorporation and eventually altered the framework through which Flynn and Coyle formulated development policy. Community groups, rather than the mayor or the BRA, first recognized that the city's spatial advantage could be the basis for a new development agenda.

The principle of bargaining with downtown developers for social benefits, a central tenet of balanced development, resulted from a 15-housing struggle in the South End, a neighborhood bordering the downtown and long a site for BRA gentrification plans (Dreier & Ehrlich, 1991). The Boston Jobs Plan, mandating the hiring of city residents on projects using public funds or subject to public regulation and made law in 1980, was first advocated by African American community organizers (King, 1981). Linkage was proposed by a newspaper columnist with ties to neighborhood groups before it was enacted in 1983 (Dreier & Ehrlich, 1991). Community groups also changed the structure of government. Hoping to bolster democratic politics, they persuaded the voters to reestablish district representation in 1981 and to endorse a nonbinding referendum establishing elected neighborhood councils in 1983.

Flynn, a neighborhood mayor, thus had the good fortune to preside over an economic boom with new powers to share that boom with his constituencies. A network of new and old city agencies delivered the benefits of housing units, jobs, job training programs, and neighborhood public works projects, demonstrating Flynn's commitments. But a distinct unwillingness to share power with community organizations, particularly in poor and potentially marketable neighborhoods, was still evident under Flynn. In Flynn's later years, community groups across the city complained that the mayor too often sided with institutions and developers, and they attacked Flynn's refusal to provide neighborhood councils with financial resources, veto power, or a permanent status (Devine, 1989). The debate over neighborhood councils takes on new importance when we realize that Boston's residents stopped voting; turnout in mayoral elections plunged by 50% (100,000 votes) during Flynn's years in office (*Boston City Document No. 10*). In the face of a weak city council, popular representation still depended on informal, bureaucratic procedures defined by the mayor.

I would argue, however, that the limited achievements of Boston's grassroots activism cannot be completely explained by the dynamics of economic restructuring, institutional arrangements, and mayoral agendas. Plotkin's notion of "enclave consciousness"—the defense of race, class, and neighborhood—seems a particularly apt way of understanding both the vitality and the limits of Boston's community politics (Plotkin, 1990). Boston's community groups have remained most committed to controlling development in their own neighborhoods: Flynn's policies thus met the agendas of many organizations even as they angered others (AuCoin, 1991). The significant victories of the Boston Jobs Plan and linkage were less the result of citywide, multineighborhood organizing pressures than of Kevin White's desire to win re-election by appealing to the concerns of a few, mobilized communities.

The major stumbling block to building broader political movements is the city's bitter racial politics. Despite the long-standing and widespread opposition to the BRA's growth strategies in poor neighborhoods of all races, racial hostilities are peculiarly implicated in community development struggles. With few exceptions, the most visible and radical challenges have been organized in the city's multiracial and African American neighborhoods—often by black political leaders. Such alternatives have faced the added difficulty of overcoming community racism as well as the resistance of mayors, the BRA, and business elites.

The governing coalition was thus only partially successful at realizing its community participation agenda. None of Boston's mayors devised a lasting set of arrangements, and thus, community groups have continued to demand mayoral attention. At the same time, sustained activism has achieved only a few lasting victories. In contrast with business agendas still privileged despite a decline in business mobilization, the demands of poor neighborhoods have typically been reshaped by state actors into policies more conducive to their interests.

# Conclusion

This chapter has demonstrated how investigating the *interconnections* between economic growth and state reorganization enhances our understanding of postwar development policy making. I emphasize the word *interconnections* because it is difficult to disentangle economic from political restructuring and vice versa. The Boston case illustrates that reworking the institutions of the local state can be crucial to the construction of spatial advantage and to the survival of a governing coalition. Moreover, the politics of state reform cannot be studied without an appreciation of economic context or market: The success of institutional reform depended on processes of economic restructuring linked to regional dynamics beyond the city limits. Those economic processes were susceptible to local decisions because key firms in certain economic sectors desired to remain or locate in the city.

In this example, I would argue that local governance is best examined as a complex effort at addressing different but interconnected political and economic dynamics, much like its formulation in regulation theory. Yet governance is highly political in the regime theory sense: Governance involved hard negotiations and compromises between actors with different agendas who came to see their fates as linked and continued to work together more or less closely for almost 50 years. Whether or not seven years of recession has put an end to this governance strategy, or whether, as in the 1970s, a new round of economic growth will sustain its existence, is a major issue facing Boston's elites and its many struggling communities.

## Notes

1. This shift in focus is especially clear in the changing analysis of business power. Rather than examining how business's inclusion in the governing coalition relates to its structural position in the local political economy, this scholarship emphasizes the organizational resources business brings to bear on urban governance problems.

2. Noyelle and Stanback also include Cleveland in this group. Cleveland was excluded from this comparison because of data constraints. Employment data are derived from the U.S. Bureau of the Census *County Business Patterns.* This series reports for St. Louis and Baltimore as independent cities; the City of Philadelphia is coterminous with the county; more than 96% of the employment in Suffolk County is located within Boston. In the Cleveland case, Cuyahoga County includes jurisdictions and employment beyond municipal limits.

3. Total Employment (excludes government and railroad workers and the self-employed)

|              | 1951    | 1962    | 1970    | 1980    | 1990    |
|--------------|---------|---------|---------|---------|---------|
| Baltimore    | 344,446 | 340,303 | 367,249 | 308,422 | 311,161 |
| Boston       | 399,901 | 421,765 | 466,985 | 440,543 | 510,072 |
| Philadelphia | 777,662 | 732,347 | 775,503 | 632,080 | 612,419 |
| St. Louis    | 419,813 | 355,739 | 376,113 | 286,896 | 257,531 |

SOURCE: U.S. Bureau of the Census, *County Business Patterns.*
NOTE: Data are published only for counties; see note 2.

4. In this chapter, I use the term *business services* as defined by the U.S. census. By *producer services,* I am denoting a broader range of economic activities that could be included as business services in addition to those so categorized by the census. The term *producer services* also includes the census's categories of finance, insurance, and real estate; legal services; and administration and miscellaneous services (architecture, engineering, accounting). If we were to compare the four cities on business services alone, Boston would still rank first.

5. In 1951, Boston contained 46% of all nongovernment jobs in the SMSA compared with Baltimore's 81% share of all SMSA nongovernment employment, Philadelphia's 71.5% share, and St. Louis's 72.5% share.

6. Other city agencies, much smaller in personnel and budget, pursued slightly different agendas: The Economic Development and Industrial Commission worked to preserve manufacturing employment; the Public Facilities Department and the Mayor's Office and Neighborhood Services worked with community groups on local infrastructural projects, housing, and some small business development.

# 9

# City Planning and the Postwar Regime in Philadelphia

## ROBERT A. BEAUREGARD

I n the United States, the now-broken postwar social contract between capital and labor neither burst forth all at once nor spread evenly across the political landscape. Although national-level actors—corporate capital, labor unions, political parties, and the federal government—struck an accord that enabled the welfare state to expand relatively unchallenged, promised peaceful labor relations in return for a family wage, and committed the state and capital to a politics of growth (Wolfe, 1981), local political arrangements also had to be forged to extend this new mode of social regulation to the cities. In the late 1940s and early 1950s, however, local elites were as likely to resist as to embrace public-private initiatives whose goal was to manage the economic and social problems of the city.

The purpose of this chapter is to explore the postwar ascendance of Philadelphia's developmental regime from the perspective of city planning, a local

AUTHOR'S NOTE: My gratitude is extended to Mark Goodwin, Susan Fainstein, Andrew Jonas, Mickey Lauria, and Jon Pierre, all of whom made helpful comments on an earlier draft of this chapter.

governmental activity that reinforced and helped diffuse the values and arrangements of the postwar social contract. Starting in the early 1940s and extending
through to the early 1960s, a progrowth coalition of political, economic, and
civic leaders and organizations attempted to return Philadelphia to its once-
dominant regional status. Initially opposed by entrenched business interests and
a contented political machine, the reformers eventually triumphed. Central to
their efforts was a planning ideology and a planning apparatus, the Philadelphia
City Planning Commission, along with an operating agency, the Philadelphia Redevelopment Authority. The former generated ideas to justify and images
to guide numerous large-scale development projects, and the latter used federal,
state, and local legislation and funding to transform visionary plans into local
action.

   Philadelphia was not unique. In numerous cities across the country, reform
groups addressed local problems generated by decades of disinvestment and the
flight of industries and households: falling real estate values, slums, declining
job opportunities, and an expanding class of poor and minority residents (Lowe,
1967). Most often, publicly initiated local development was the solution of
choice. Reinforced by an admittedly fragile, national postwar consensus, urban
developmental regimes proliferated and a new mode of social regulation was
solidified locally.

   The institutionalization of elements of the new mode of social regulation
across a large number of cities and regions enabled that mode to survive at the
national level. From a theoretical perspective, neither regime theory nor regulation theory alone suffices to explain this phenomenon. A conceptual middle
ground has to be prepared.

## A Note on Theory

   The intent of the case study is to illustrate the limitations of both regime
theory and regulation theory in explaining the formation of local developmental
regimes. On the one hand, local regimes do not flow automatically from the
fiscal constraints of local government, the motivations of land-based elites, and
the threats posed by mobile capital. Such regimes are more contingent than
regime theory implies and more complex in their formation. On the other hand,
a national mode of social regulation committed to a politics of growth is a
necessary but not sufficient condition for regime formation.

   Regime theory, of course, takes as its object the informal political arrangements among private and public actors that center on economic and, less so,

social development in cities and regions (Elkin, 1987; Kantor, 1988; Sanders & Stone, 1987; Stone, 1989). These arrangements emerge because local governments alone lack the resources and capabilities to engender economic growth, because private sector actors (unable to capture all or most of the benefits) will not act alone, because the private sector requires public goods (such as airports) to thrive, because the intergovernmental system makes local government rely on own-source revenues, and because informal arrangements provide greater flexibility, greater distance from public accountability, and a greater ability to distribute the benefits of development to those who constitute the regime. Although regime theorists (Fainstein & Fainstein, 1983; Mollenkopf, 1983) recognize that not all urban areas will develop progrowth coalitions, and that some coalitions are stronger than others, their formation is undertheorized. The theoretical emphasis is on the perpetuation of existing regimes.

Furthermore, regime theory takes contextuating factors as given. Too little attention is paid to how the regime is connected to structures and processes operating at larger geopolitical scales (particularly the mode of social regulation). Moreover, urban regimes are not considered important contributors to the latter's functioning. Urban regimes are essentially isolated; to produce local growth, the regime must be indifferent to the consequences of its actions for other localities, regions, and developmental regimes, including the national regime. Regime theory is about the selfishness and self-centeredness of local progrowth coalitions.

In regime theory, the context is treated as impetus and then discarded. Issues of uneven spatial development, shifting intergovernmental relations, and the local consequences of actors operating at different spatial scales are all treated in a less than satisfactory manner. Admittedly, regime theorists have posed the question of functioning rather than formation. Nevertheless, the regime's performance is not solely a matter of the internal distribution of benefits, consensus around development projects, and relations with the governing coalition. Urban regimes are neither isolated nor closed arrangements.

Regulation theorists are only beginning to address local government (Goodwin, Duncan, & Halford, 1993; Stoker, 1990) and then only to recognize the impact of shifts in modes of social regulation on local government, not to theorize how local developmental regimes enable, and are enabled by, national modes of social regulation. Consequently, the dynamics of institution building at various spatial scales and how this is linked to the full flowering of a mode of social regulation is undertheorized. The goal of regulation theory, to the contrary, is to explain the regulatory mechanisms, cast as institutional relations and societal conditions, that enable a particular regime of accumulation to function (Brenner & Glick, 1991; Stoker, 1990). Capitalism, in its various guises, is the regime of accumulation and the threats to its performance, drawing weakly on Marxist theory (Walker, 1995), take the form of crises engendered by internal contradictions. Lacking cultural values, social arrangements, and political institutions that reinforce the workings and values of the regime of accumulation—everything from a educational system that instills proper work habits to state banking systems that regulate the money supply—the regime will perform poorly, experience debilitating crises, and face arrested development.

Introducing space into the discussion, and specifically spatial scales and their political divisions, makes it apparent that a regime of accumulation, if it is to be any more than an "enclave economy," has to be supported by a mode of social regulation that extends geographically throughout the territory in which the regime of accumulation is attempting to operate. Institutions and conditions compatible with the corresponding mode of social regulation must be manifested at a variety of spatial scales within that territory. The postwar social contract in the United States worked to the extent that it was reflected in state and local forms. Developmental urban regimes translated the postwar social contract into more or less compatible local policies.

Regulation theory, however, only weakly addresses this side of its argument. Although regulation theory is quite sensitive to national differences in the mode of social regulation, thereby implying spatial variation, it is relatively inattentive to subnational elements of the mode of social regulation. The nation is the geopolitical unit of analysis for regulation theory, a bias that is less and less tenable in the face of regional economic integration and transnational corporations. Both cross-national and subnational variations need to be considered.

The theorization of the formation and perpetuation of subnational elements, however, must avoid both a simple historical sequencing and a strict spatial correspondence. It is unlikely that a national mode of social regulation would emerge and diffuse simultaneously throughout the nation. More than likely, elements of the mode of social regulation appear out of sequence at different spatial scales. Consequently, the expectation is of an uneven spatial development of the mode of social regulation, with some localities creating supportive institutional arrangements, others struggling to do so, still others resisting, and a few wholly indifferent.

Although a single case study cannot explain all these undertheorized aspects of regime theory and regulation theory, it can illustrate the issues and suggest a direction of inquiry. The rise of city planning and its encompassing progrowth coalition in Philadelphia, a city that created one of the earliest and most fully realized developmental regimes, is a good place to begin.

# City Planning and Urban Reform

From Philadelphia's very beginning, city planning has been a presence. As part of William Penn's decision in 1681 to settle the land chartered to him by the King of England, Penn hired Thomas Holme, a surveyor, to lay out the lots that would be purchased by settlers. From then until the early 1940s, city planning never wholly disappeared from the public debate about how Philadelphia should grow (Cutler, 1980, pp. 257-264; Philadelphia City Planning Commission, 1954).

For example, in 1721 Philadelphia was granted permission by the Pennsylvania Assembly to appoint "surveyors and regulators" whose responsibility would be to ensure "orderly street and building lines" (Cutler, 1980, p. 257). When the city consolidated with surrounding communities in 1854, a Board of Surveyors and Regulators was created. Its successor attempted to take charge of development issues in 1885 and eight years later established a general plans division to oversee streets and traffic.

The Permanent Committee on City Plans (1912-1919) was set up to develop comprehensive land use plans, and in 1919 a zoning commission was written into the charter. The zoning commission languished, as did a City Council proposal for a planning commission in 1929. Neither planning nor zoning was prominent during the Depression of the 1930s.

Consequently, when civic leaders began to debate Philadelphia's postwar prospects in the early 1940s, they did so against a long history of attempts to regulate the city's growth. The issue was the establishment of a permanent planning commission that would strengthen the government's commitment to city planning and address the need for housing improvement. The provision of adequate streets and legal property lines had become a minor concern. High on planning's agenda were vast areas of slums and blight, a serious housing shortage, and a shrinking economic base brought on by industrial decentralization (Beauregard, 1989b, pp. 198-204).

## Background

In the early 1940s, Philadelphia suffered from urban problems common to older, industrial cities of the Northeast and Midwest (Vigman, 1955, pp. 112-118). Over the past decade, the housing stock had deteriorated and little new housing had been built. A housing shortage ensued and slums spread. The Philadelphia Housing Authority in 1944 claimed that 15% of the occupied residential units in the city were substandard (Bauman, 1981, p. 2). Furthermore, the lack of new commercial construction had eroded the tax base and the downtown area was faced with declining property values. Overall, the assessed value of taxable property in current dollars dropped from $3.4 billion in 1929 to $2.5 billion in 1939 (Chamber of Commerce, 1951, p. 58). By 1939, the debt service at 39% was the largest single charge against the city's budget and per capita expenditures on municipal services were well below that in New York, Boston, and Detroit (Vigman, 1955, pp. 133-134).

In addition, industries and households were leaving the city, a long-running decentralization that had been slowed by the Depression but was now accelerating (Clark & Clark, 1982, p. 659). Moving into the city were Negroes—the contemporary label—from the South; from 1930 to 1940, the nonwhite population grew by 14% and the white population fell by 2%. Negroes often ended up in the poorest neighborhoods and arrived when manufacturing jobs were shrinking as a proportion of all city jobs.

Regionally, Philadelphia was losing its demographic, economic, and social dominance; the city's population was essentially unchanged from 1930 to 1940, rose slightly in 1950, and then began a steady decline. Its share of the metropolitan population dropped from 61% in 1940 to 56% in 1950. Local leaders lacked the prescience to predict the massive suburbanization that would soon follow, but they recognized that Philadelphia's problems demanded their attention.

Not all of Philadelphia's leaders were convinced that something should and could be done, however. The city government had been under the control of a Republican political machine for some time, and even though Republican mayors in the early 1940s supported the establishment of a City Planning Commission, their intent was never to create an activist local government that would combine public and private investment in development projects, take responsibility for housing, and engage in a broad range of social programs. Under the Republican machine the city government followed conservative business practices; it spent little and kept taxes low. An activist government would be a significant departure from current practices.

During the Depression, for example, Mayor J. Hampton Moore had turned aside a federal loan to provide improvements to the municipal airport and in 1934 once again rejected the use of federal funds for public works projects. From 1932 to 1937, the city government had contributed nothing to direct relief. Only the replacement of Mayor Moore in 1936 by a less rigid Republican mayor led to Philadelphia accepting funding from the Works Progress Administration (Tinkcom, 1982, pp. 611-613).

Then, it was possible to believe that with the end of the war the city would return to the prosperity of the 1920s. Not until government mismanagement generated a national image of incompetence and led to serious fiscal problems that posed a potential threat to the business community did economic elites consider reform.

## City Planning Commission

Against this background, civic leaders began to lobby for an ordinance to create a City Planning Commission (CPC) (Bacon, 1943; Petshek, 1973, pp. 17-21). They included "civic-minded technicians" from the City Policy Committee, representatives from the Philadelphia Housing Association (an organization whose origins were in Progressive housing reform), the Junior Board of Commerce, and the Lawyer's Council on Civic Affairs.

The City Policy Committee comprised young business leaders (known locally as the "Young Turks") with a "penchant for introducing new approaches" (Bauman, 1987, p. 97; Petshek, 1973, p. 19). They came together in 1939 to encourage the local government to become more involved in the city's future. Spurred on by the president of the University of Pennsylvania, Thomas Gates, and a civic-minded lawyer, Walter P. Phillips, these groups formed the Joint Committee on City Planning and worked with the Bureau of Municipal Research to prepare reports and recommendations regarding the proposed CPC.

To solidify support for city planning, the Philadelphia Housing Association convinced the American City Planning Institute to hold its national conference in Philadelphia. At the 1941 conference, a representative of the Joint Committee spoke publicly of the need for a City Planning Commission. Then-mayor Robert E. Lamberton heard the speech and delegated to the Joint Committee the task of formulating specific proposals. Lamberton, though, "died the very day he received the report of the Joint Committee" and "his successor failed to see any need for city planning and refused to take any action" (Petshek, 1973, p. 20). In April 1942, the "Ordinance to Create a City Planning Commission" was introduced to the City Council but lacked strong Council support.

At public hearings in July, 30 civic groups testified. In September, 50 organizations joined to form the Action Committee on City Planning so as to bring additional pressure on the city council to pass the ordinance. The overall effort was widely representative with "major professional societies, business and labor groups, women's organizations, social agencies, and neighborhood associations from all parts of the city" (Bacon, 1943, p. 62). (The Chamber of Commerce and the Real Estate Board, two major actors in the business community, did not join the Action Committee but remained independent of the reformers.)

The Action Committee subsequently met with Mayor Bernard Samuel, who assigned the Director of Public Works to work with it to detail the proposal. In early December 1942, a final hearing was held before the Planning Committee of Council, a hearing attended by 60 civic organizations. Council members were greatly impressed by the testimony of Edward Hopkinson, a senior partner in the investment firm of Drexel and Company, chairman of the executive committee of the Chamber of Commerce, and a prominent member of the "Old Philadelphia" Establishment. The City Planning Commission, Hopkinson believed, would enable the city to undertake long-neglected capital projects and improve the city's credit rating. The amended ordinance was adopted by Council on December 10 and the official 1943 budget included $40,000 to fund the CPC. Hopkinson was named commission chairman, and Robert B. Mitchell, the former head of the Urban Section of the New Deal's National Resources Planning Board, was appointed the first executive director.

Although the establishment of a City Planning Commission might have been viewed by those involved as "a dramatic story of successful cooperation between public officials and citizens' groups" (Bacon, 1943, p. 62), that success did not directly and immediately translate into a city government actively involved in stimulating private sector development. The Republican machine was still in office, the charter (reformers believed) hindered a development-oriented government, and the business community was not wholly supportive.

## National Forces and Debates

Local debates about the need for a City Planning Commission and a development-oriented local government were framed by a national discussion about the deteriorated conditions of the industrial cities (Beauregard, 1993, pp. 74-105) and what governmental response would be most appropriate (Bauman, 1981; Gelfand, 1975, pp. 105-156). That discussion had a number of intersecting dimensions.

One of those dimensions involved the desirability of federal government (and particularly national or federal government) intervention in urban redevelopment. The alternative was "planless growth" that would rely on private investors to reinvest (or not) in the central cities. Lobbying for national intervention were the housers and planners who called for a major commitment of resources to comprehensive planning and redevelopment. Opposed to federal governmental involvement were such organizations as the Twentieth Century Fund and the

National Association of Real Estate Boards, which held that government intervention would undermine the private investment market.

Eventually, this debate was resolved (temporarily, as we now know) in favor of those who supported a more activist federal government. In its urban guise, the resolution encompassed the Urban Renewal program of the 1949 and 1954 Housing Acts, the public housing program of the 1937 Housing Act, and the federal home mortgage insurance program that originated with the 1934 Housing Act (Gelfand, 1975).

Along another dimension, the discussion considered what seemed to be diametrically opposed choices: whether to rehouse slum dwellers or reconstruct the downtowns. Those who argued for the latter did so by focusing on the regional economic importance of the urban core, whereas those arguing for the former were more concerned with the social needs of the city's low-income residents and related racial tensions. For them the issue was not the city's regional dominance but the desirability of the city as a place to live.

In the 1940s, the debate between those committed to slum clearance and public housing and those who preferred to address blighted commercial property and central business district redevelopment constituted an emerging split between housing planners and redevelopment planners. Nevertheless, the interests joined around the need to clear slums and engage in comprehensive planning. Slums often abutted or were spatially intertwined with commercial areas, particularly in older industrial cities. Moreover, comprehensive planning would enhance implementation and ensure that efforts made sense in the larger urban and regional context.

The national debate did not pose simple choices, and local officials and civic and business leaders seldom agreed how to proceed. Which level of government might act in what capacity, the possibility of disallowing governmental intervention altogether, and the choice of whether and to what degree to emphasize housing or downtown redevelopment all had to be decided, and the decisions would influence the political and business support that would be available for undertaking large-scale development projects.

## Local Reform

In Philadelphia, these choices were made against the backdrop of local government reform, specifically as it centered on charter revision and the replacement of the Republican political machine by Democratic reformers. Public complaints about corruption and the wastefulness of government began back in the late 1930s, but not until the latter years of the 1940 decade was a concerted reform effort mounted.

Political reformers had more than waste and corruption on which to base their calls for "throwing the rascals out." The quality of water in the city was notoriously poor, city officials could not account for $40 million in city expenditures in 1948, and since the 1930s the government had verged on and entered bankruptcy only to be bailed out by the local banks. The fire marshal was convicted and sent to prison, and embezzlement and forgery were uncovered in the Department of Supplies and Purchases. A number of city employees committed suicide rather than face grand juries. Supported by the newspapers, civic groups exposed numerous instances of mismanagement and corruption, and thereby prepared the ground for overthrowing "the nation's most entrenched and incredibly corrupt political machine" (Clark & Clark, 1982, p. 652).

The first act of the reformers was the mayoral campaign of Richardson Dilworth in 1947. Although Dilworth lost, he focused public attention on governmental corruption and inefficiencies. A mayoral commission, the Committee of Fifteen, created to look into employee pay raises also ended up documenting government waste. One year later, one hundred or so businessmen met to consider how they might contribute to the reform movement. The outcome of their meeting was the Greater Philadelphia Movement (GPM), key local corporate executives dedicated to raising public awareness of the need for specific citywide projects (Petshek, 1973, pp. 26-31).

The reform leadership and GPM joined forces in 1949 to obtain state legislation that would create a Philadelphia Home Rule City Charter Commission. Business support grew out of a realization that municipal government that wasted tax dollars and provided substandard public services would eventually hinder local business performance. Subsequently, the new city charter was voted on and passed in a special election in April of 1951. It created a strong-mayor government, installed a merit personnel system, provided the mayor with the power to form citizen commissions such as the City Planning Commission and the Commission on Human Relations, and streamlined the city administration. The last included creating the positions of managing director, director of finance, and city representative/director of commerce who would supervise economic development (Clark & Clark, 1982, pp. 654-657; Petshek, 1973, pp. 34-38).

The success of the charter reform provided impetus for the reform (or "independent") Democrats, led by Joseph S. Clark and Richardson Dilworth, to elect one of their own to the mayoralty. In 1949, Dilworth ran for city treasurer and Clark for controller and both won, but not until 1952 did the reform Democrats take control of the mayor's office when Clark won election and the Democrats gained a majority in the City Council. (Dilworth became the district attorney after unsuccessfully running for governor a year earlier.) Their campaigns stressed ending municipal corruption and resolving the city's debt problems.

Dilworth and Clark were committed to addressing Philadelphia's social, physical, and economic problems. It was clear to the voters, and particularly to an emerging segment of the business community—professional and real estate interests—that this meant a government that would use comprehensive planning to address slums, blight and industrial decline. "The most distinctive contribution of Philadelphia's reform administration was to embark on programs which would affect the development of the city—physically, economically and socially—and leave a permanent imprint," what came to be called Philadelphia's renaissance (Petshek, 1973, p. 53).

The city's renewal was possible only because an important element of the business community committed itself to redevelopment. The turning point was the decision by Albert M. Greenfield, a major banking and real estate executive and a key member of the Chamber of Commerce, to involve himself in the city's revitalization. His appointment as chairman of the City Planning Commission by Clark's successor, Richardson Dilworth, led to an expansion of the CPC's budget, a major increase in the city's appropriations for urban renewal, and a more aggressive use of federal urban renewal legislation (Lowe, 1967, pp. 341-343). Greenfield's support allowed Edmund Bacon, executive director of the Philadelphia City Planning Commission after Robert Mitchell, to promulgate his ideas for the redevelopment of downtown Philadelphia.

Clark served two terms before becoming a U.S. senator from Pennsylvania. Dilworth replaced him as mayor in 1956 and served almost two terms in office before resigning in 1962 to run for governor, a race that he lost (again). The acting mayor was James H. J. Tate, an "organization man" more comfortable with machine than reform politics. The independent Democrats unsuccessfully ran Walter M. Phillips against Tate in the next election. Tate's victory had implications for government-led development; his relations with the civic and business groups that had pushed for reform were strained (Petshek, 1973, pp. 81-82). Tate's mayoralty effectively severed the reform Democrats' control over municipal government.

Deepening social problems and heightened racial tensions in the city during the 1960s eventually led to the 1971 election of Frank L. Rizzo as mayor (Adams et al., 1991, pp. 126-129, 149). Rizzo, a former police chief, portrayed himself as a law-and-order candidate who would deal forcefully with radicals and racial unrest. He drew most of his support from the white working-class and opposition from liberals and blacks. A defining issue in his mayoralty was his refusal to allow a public housing project to be built in a white neighborhood. That refusal led to federal sanctions by the U.S. Department of Housing and Urban Development that denied Philadelphia access to Urban Development Action Grants, a key redevelopment tool of the late 1970s. To a great extent this change in political climate severely weakened the development-based reform coalition, although one commentator has noted that the Rizzo administration was "particularly sensitive to the necessity for economic reinvestment in Philadelphia" (Wolf, 1982, p. 710). Nevertheless, the era of large-scale redevelopment projects had passed and low-income housing was not on the Rizzo agenda.

### Redevelopment

To address the housing and economic needs of the city, ideas had to be developed and implemented. Redevelopment was the goal. Ed Bacon, executive director of the City Planning Commission and an urban designer, had grandiose ideas concerning how Center City could be transformed through renewal projects. He focused not only on commercial areas such as Market Street, where the city's major department stores were located and an area west of city hall that eventually became the famed Penn Center office building complex, but also on residential areas, particularly the now widely praised Society Hill.

The City Planning Commission, however, was not an implementation body; it could only propose and regulate. As headed by Bacon, its focus was physical design rather than the economic feasibility and social impacts of redevelopment. The CPC certified renewal areas, Bacon's staff proposed visionary plans, and zoning fell within its purview. However, Bacon's lack of access to the mayor's office and his reluctance to compromise made the CPC less influential in redevelopment than it might have been (Barnett & Miller, 1983). Instead, the Philadelphia Redevelopment Authority took center stage.

Federal legislation played a significant role in this shift. Philadelphia's Redevelopment Authority had been established in 1945 to eliminate blighted areas but did not acquire the tools to undertake large-scale purchase and clearance of redevelopment sites until the 1949 federal Housing Act put the urban renewal program in place. The planning director for the Redevelopment Authority, David Wallace, subsequently took charge of all aspects of renewal planning and was the person (rather than Bacon) who city officials felt best qualified to lead the renewal effort (Petshek, 1973, p. 100).

The federal government was also influential on the housing side. The city's Housing Authority had been established in 1937 as a direct consequence of the U.S. Housing Authorities Act of 1937 that had turned public housing into a nationwide program. When combined with the slum clearance powers of urban renewal, the Housing Authority possessed powerful tools to remove slums and replace them with quality, low-income housing (Bauman, 1987).

Once the government had dedicated itself to redevelopment, key individuals within the business community had become involved, and the appropriate public agencies were armed with the necessary tools, the City Planning Commission, but more so the Redevelopment Authority, began to influence the future form of the inner city (Adde, 1969; Beauregard, 1989a, 1989b; Cybriwsky & Western, 1982; Lowe, 1967, pp. 313-404).

Renewal of the 22-acre Penn Center site in 1952 was the first project undertaken. It required the purchase and removal of a railroad station and its "Chinese Wall" of railroad tracks, and resulted in the construction of a cluster of office buildings on a plaza above a complex of shops and transit stations. In that same year, the CPC also proposed clearance and redevelopment of an area adjacent to Independence Hall where the U.S. Constitution had been written and signed. The densely built blocks of industrial and commercial buildings in front of the hall were demolished and the cleared land made into a national park. New office buildings were also constructed along one edge of the site.

The Society Hill project was started soon thereafter. A wholesale food market was replaced by high-rise apartment buildings and an adjacent area of deteriorated colonial-era row houses was preserved and rehabilitated for middle-income homeowners. The mayor and other civic and business leaders were determined to keep the middle class in the city and to reorient the city's economy to business and professional services located in the downtown core.

In the neighborhoods, urban renewal involved slum clearance and the construction of government-subsidized housing. Here, the City Planning Commission worked with both the Redevelopment Authority and the Housing Authority to clear slums and construct public housing, though not without (in later years) resistance (Bauman, 1987).

By many measures, these projects were successes and the ideas and actions of planners, whether in the CPC or the Redevelopment Authority, were essential in justifying and guiding redevelopment.

Overall, city planning in Philadelphia benefited from a government reform movement, the difficult problems faced by a declining industrial city, business community support, and an array of federal policy tools that gave legitimacy and function to planning ideas (Bauman, 1987, pp. 90-93). Under these conditions, planners were able to shape central business district redevelopment and residential renewal. By the 1970s, a combination of racial discord, opposition to large-scale slum clearance, and diminished federal support for redevelopment weakened planning's influence (Cybriwsky & Western, 1982, pp. 358-361).

## Regimes and Regulation

The period from 1942 to 1962 witnessed the formation and demise of Philadelphia's developmental regime. The regime's fortunes were not wholly a result of local motivations, capacities, and cooperation; these were necessary but not sufficient conditions. Although locally contingent, the regime was made possible by a national shift in the mode of regulation. Without that, the regime would not have achieved the successes attributed to it. When that mode of regulation began to break down, the consequences reverberated to Philadelphia and its development-oriented regime unraveled. The objective conditions signaling the need for reinvestment worsened, but large-scale, publicly led reinvestment diminished. This connection between the local and the contextual and between regime theory and regulation theory is what makes this case theoretically interesting.

The period from the early 1940s to the early 1960s in Philadelphia is familiar to regime theorists (Fainstein & Fainstein, 1983; Kantor, 1988; Stone, 1989). Business and political leaders came together to engage government in publicly subsidized development, governing regimes were formed that differed from electoral coalitions, and city leaders debated the image of the city and its niche within the region (Pagano & Bowman, 1995, pp. 44-67).

To change the direction of Philadelphia's government, the Republican political machine had to be supplanted by a public-regarding, activist government. The "independent" Democrats, with the help of civic leaders and members of the business community, took on this task. These elite reformers were able to oust the Republicans both because of the scandals, corruption, and mismanagement and because disinvestment and decentralization compelled a more aggressive approach to local public and private investment.

The chosen solution was redevelopment, large-scale projects that transformed the image of the city and generated a host of investment opportunities (Stone, 1989, pp. 234-245). With the assistance of various federal programs (particularly Urban Renewal), the guidance of planners from the CPC and the Urban Redevelopment Authority, and the commitment and support of the business and financial community, major projects were implemented. Redevelopment generated a host of benefits for political leaders, the local real estate industry, and corporate business interests. Moreover, it eliminated slums and blight and improved the local infrastructure. These projects maintained the governing regime and, at least until the early 1960s, generated political support for reform.

The actions of the reformers and planners, though, were not designed to forge a governing regime but to alleviate the twin evils of municipal corruption and urban decline. The Republican political machine was viewed as corrupt because it distributed public benefits to supporters and business interests in return for votes. It was patronage driven. Reform government, on the other hand, would ensure that the benefits of publicly subsidized private investment were distributed across the whole community. If government were to take more responsibility for its citizens, patronage politics would have to be jettisoned.

Reinforcing this conclusion was the poor state of the city's central business district (CBD), industrial areas, and low-income neighborhoods. Years of disinvestment and decentralization had made Philadelphia a less attractive place in which to invest and live. City leaders were worried about losing the middle class and losing out on postwar prosperity. They saw that they would have to compete with the suburbs for new investment. They would have to not only transform the CBD but also eliminate adjacent slums. Both objective conditions and a poor public image provided the impetus for engaging in large-scale redevelopment.

The regime stayed in power through the terms of Mayors Clark and Dilworth because these arguments made sense and because the benefits of publicly subsidized development accrued to key political and economic leaders. A number of factors, though, began to drive wedges between the governing regime and electoral coalition, and this subsequently led to the regime's demise.

To explain why a development-oriented regime became established and why it subsequently receded, however, we have to augment regime theory (Fainstein, 1996). Regime theory's thrust is internal to the city itself, focusing on who cooperates, how cooperation is achieved, and for what ends (Stone, 1989, pp. 6-9). Extralocal factors are included only to frame the regime's commitment to intergovernmental competition for private investment, a commitment compelled by federal tax policy and the relative mobilities and local dependencies of capital and community.

## Modes of Regulation

The postwar urban regime was not unique to Philadelphia. Other older, industrial cities—Boston, Pittsburgh—experienced similar transformations. National conditions made this possible, but when national conditions changed, these regimes faltered (Fainstein & Fainstein, 1983).

Probably the most important of these national forces was the postwar consensus between capital and labor that provided the foundation for greater governmental involvement in social welfare activities. The origins of these various programs (such as public housing) can be traced to the New Deal of the 1930s and before, but the national commitment took place after the war. The prosperity of the 1950s and 1960s was built not only on labor-management peace and the family wage but also on the federal government's commitment to macroeconomic pump priming and a variety of compensatory programs for those outside the mainstream of prosperity. The ideological thrust was to legitimize activist government and a concern for the poor. The transition to a Keynesian welfare state based on a fordist accumulation regime enabled national political regimes to fortify themselves locally (Mollenkopf, 1983). The local state became both an object and agent of this mode of social regulation. Its charge was the uneven spatial development of capitalism, an issue addressed through slum clearance, urban renewal, and public housing.

Urban decline, of course, was not localized in Philadelphia. The loss of population and industry was part of a larger industrial restructuring—a breakdown of fordism—that eventually downsized manufacturing employment (a sector on which these older industrial cities had depended) and of changes in transportation technology that facilitated decentralization. In addition, war mobilization had "opened up" the west and south to development, and this set in motion regional shifts of capital and population that further undermined cities such as Philadelphia.

Large-scale, local redevelopment was made possible not just by national legislation and a national debate about housing and redevelopment but also by changes in the real estate industry. An infrastructure supportive of big development projects had emerged just before and during the war. It included not only large developers but also large construction companies and financial institutions willing to fund such projects. Moreover, experts (planners and policy analysts) appeared who understood how to do such projects, providing key knowledge regarding everything from eminent domain to site clearance to urban design.

In effect, the United States had shifted from a mode of regulation based on a relatively passive federal government disconnected from a cluster of regional economies to one where mass production and consumption occurred in national and even international markets, the federal government took on welfare obligations, national and subnational governments agreed to support private investment and engage in a politics of growth, and capital suspended its attacks on labor. Urban regimes evolved within and contributed to this structured coherence.

The postwar mode of social regulation was short-lived. Its life span at best extended only from the building boom of the 1920s to the recession of the early 1970s. Industrial restructuring weakened the capital-labor accord that had been forged in the mass production unionized industries, and the flight from industrial cities and regions undermined the political support for a large welfare state. Thus the federal government began its withdrawal, and activist governments at all levels began to lose legitimacy. Subsequently, the Keynesian-fordist welfare regime was supplanted by a neoclassical, postfordist workfare regime.

Regardless, local city governments in certain regions were still faced with problems of decline and still in need of larger tax bases. Decentralization continued to weaken the local ties of local corporations and real estate interests. The portfolios of these business increasingly included suburban investments, and the nationalization of these firms undermined local business leadership. It was not so easy for city governments to withdraw from growth-oriented developmental politics. At the same time, middle-class supporters of reform continued to "exit" rather than "voice," and this further weakened developmental regimes.

Making development-oriented politics even more problematic in the middle to late 1960s was opposition to large-scale redevelopment, particularly as it came from the black community. Rebellion not only made redevelopment more difficult in Philadelphia but also weakened the electoral regime. Racial unrest created a backlash that allowed Frank Rizzo to become mayor, a mayor who did not have the same commitment to reform and redevelopment as Clark or Dilworth.

The opposition of the black community, the out-migration of the white middle class, and the in-migration of rural blacks were national forces that took particular form in Philadelphia. In addition, several specifically local factors subverted the growth-oriented urban regime. The first factor was the political aspirations of the two key Democratic reformers: Clark and Dilworth. Clark moved on to become a U.S. senator, and Dilworth was always running for higher office; no person with similar reform credentials followed them. The second factor was the lingering remnants of machine politics. The "independent" Democrats never wholly took control of the local Democratic Party. Thus, when the local coalition weakened, Mayor Tate was able to win election. Although he was not opposed to the reforms and development projects of the reform mayors, neither did he enthusiastically embrace an entrepreneurial approach to local government. Opposition to redevelopment projects likely reinforced his inclinations.

Through the time of Philadelphia's urban regime, planning played an important role in providing ideas and images and in managing redevelopment itself. Planners were useful, although minor, players. The development-oriented regime supported and relied on a particular variant of comprehensive planning, one that fell out of the favor both publicly and within the planning profession in the 1960s. When the regime no longer offered its support, formal planning in Philadelphia became less heroic.

## Conclusion

To tell the full story of city planning and its development regime in Philadelphia, we must escape the limited focus of regime theory and build on emerging ideas in regulation theory. Regime theory operates with a strong emphasis on local politics. We learn mainly why and how business, civic and political leaders come together to manage public and private investment in large-scale projects, stay together to do so, and shape public opinion. Although regime theory answers these questions well, it does not adequately probe the conditions that make urban developmental regimes possible or connect the collective actions in these localities to a national mode of social regulation.

Regulation theory's interest in exploring the institutional structures that support different regimes of accumulation and the recent placement of local government within this framework may provide answers to questions not addressed by regime theory. In the absence of a better understanding about how modes of social regulation operate across space, however, this will not happen. Whether alone or together, and as currently formulated, regime theory and regulation theory are insufficient for understanding the spatial and social robustness of the capitalist political economy.

# 10

Cleveland: The "Comeback" City

*The Politics of Redevelopment
and Sports Stadiums Amidst
Urban Decline*

W. DENNIS KEATING

leveland, Ohio, has called itself and been hailed as "the comeback city" (Peterson, 1995). This reflects the economic, social, and population decline of this "rustbelt" city from the 1950s through the 1980s. The burning of the polluted Cuyahoga River in 1969, celebrated in song and jokes, made Cleveland into the "mistake by the lake."

Major downtown redevelopment projects such as office, retail and hotel developments, the renaissance of Cleveland's Flats (formerly the heart of its port and industrial valley), and such "blockbuster" projects as the Gateway sports complex opened in 1994 and the Rock and Roll Hall of Fame opened in 1995, both to national acclaim, contributed to Cleveland's image as a revitalized city. The city's media, elected officials, civic and business leaders (Cleveland Tomorrow), chamber of commerce (Greater Cleveland Growth Association), tourist and convention bureau, and the "New Cleveland" marketing campaign have cultivated this image of success.

In this chapter, this revitalized city claim will be challenged. This critique will analyze the politics of Cleveland's modern redevelopment efforts, focusing on downtown sports complexes. Public financing has been critical to the public-private partnerships that have been hailed so often as indicators of Cleveland's revitalization success. Their creation and use will be critically examined, as well as their impacts on the city's residents. Several models will be referred to in this case study: the city as a growth machine, urban political regimes, regulatory theory, uneven development, and the dual city phenomenon.

The intercity competition for professional sports franchises is but one symptom of regional, national, and international economic competition. In a few cases, U.S. cities have attempted to use sports as a major vehicle for economic development, including tourism, for example Indianapolis (Rosentraub, Swindell, Prezybylski, & Mullins, 1994) and Atlanta, which hosted the 1996 Summer Olympics. The overriding significance of attracting these sports franchises is that it is felt that this maintains or enhances the city's status as a competitor for growth and its image as a top-tier city.

The competition for professional sports franchises raises the urban regime theory explanation of business domination of or alliance with local elected officials in providing the mostly public financial incentives and facilities demanded by sports franchises. In the United States, these franchises are privately, not publicly, owned, quite often by extremely wealthy investors. Increasingly, the owners are not tied to the city or region in which the team is located, resulting in mobile franchises in search of higher profits. This means that cities have been forced to make stadium lease arrangements in which the private owners take the least economic risk and reap most of the financial rewards. The host cities, often the owners of the facilities in which the teams play, try to recover the operating costs of publicly owned stadia and otherwise hope for indirect economic benefits from sporting events. Baade (1996), Baade and Dye (1988), Baim (1992), and Rosentraub et al. (1994) argue that these economic benefits are inconsequential.

Nationally, sports franchises are monopolistic cartels governed by the economic self-interest of the owners, individually and collectively. Professional baseball has long had a unique antitrust exemption (Euchner, 1993, p. 38). Lacking such an explicit exemption, other leagues profess an inability to prevent existing franchises from moving. In the case of the National Football League (NFL), its lawsuit to prevent the Oakland Raiders relocating to Los Angeles in 1980 not only failed but cost the NFL $50 million in legal costs (Euchner, 1993). These leagues still determine whether new franchises are to be created. Attempts to persuade the U.S. Congress to intervene legislatively to regulate the movement of professional sports franchises or extend the antitrust exemption to other sports have failed. In fact, cities publicly financing new sports facilities used federally tax exempt municipal bonds for much of the costs, until the federal

Tax Reform Act of 1986 ended this provision. Thus, local governing regimes compete without intervention by the national government.

In this case study of Cleveland, the conflict between competing public and private interests is illustrated by the building of a new downtown complex for the baseball and basketball teams.

## Cleveland: Decline and Attempts at Revitalization

At the peak of its growth around the turn of the century, Cleveland ranked as the sixth largest city in the United States and was a national center of industrial and commercial development. Its economic growth stimulated its population growth, fueled by immigration from abroad (Miggins, 1995; Miller & Wheeler, 1995). In the wake of two world wars and the Great Depression, Cleveland, like other rustbelt industrial cities, saw itself poised for continued growth. Instead, with a population of 914,000 in 1950, Cleveland's population was decimated by suburbanization, deindustrialization, and racial conflicts, including race riots in 1966 and 1968 and the continuing battle over a 1977 racial desegregation order imposed by a federal judge on the Cleveland public school system (Bier, 1995). By 1990, Cleveland's population had fallen to 505,000.

Deindustrialization, beginning in the 1960s, accelerating in the 1970s, and continuing through the 1990s, left Cleveland with a much reduced industrial employment base, large tracts of vacant (and usually polluted) industrial land, tens of thousands of abandoned houses, stores, and public facilities—often symbolized by empty lots after loss of the buildings through demolition or arson—and a decaying infrastructure. Increasing white collar employment in the downtown only somewhat offset this loss (Hill, 1995). Businesses fled the city for its suburbs, other parts of the United States, and overseas in response to the city's physical and social decline and often in response to financial induce-ments such as tax abatements offered by competing jurisdictions.

To counter this phenomenon, Cleveland, like its counterparts, mounted in the 1960s one of the most ambitious urban renewal programs in the United States, focused on redevelopment of the downtown core and its immediate surround-ings. This, together with construction of a network of urban freeways, still failed to stem the flight of business and employment. In the 1980s, Cleveland was one of the most successful recipients of federal Urban Development Action Grants (UDAGs), the successor to urban renewal until its termination by Congress in 1986. Emphasizing public-private partnerships, Cleveland focused on down-town commercial redevelopment, rather than industrial or neighborhood-based projects (Keating, Krumholz, & Metzger, 1989; Keating, Krumholz, & Perry,

1995). Cleveland's downtown revitalization strategy—aiming to promote large-scale retail, office and hotel development, entertainment and sports attractions to attract tourists, and the physical modernization of the central business district—was detailed in the city's Year 2000 Downtown Plan, part of its "Civic Vision" (Cleveland City Planning Commission, 1988). Cleveland aims to be a major regional and national economic competitor.

In 1991, Cleveland published a companion citywide plan, emphasizing neighborhood commercial, industrial, and housing development (Cleveland City Planning Commission, 1991). The city did support neighborhood revitalization efforts, using the federal Community Development Block Grant (CDBG) Program created in 1974 and supporting the efforts of community development corporations, most notably the Cleveland Housing Network. Beginning with urban renewal, then UDAG, and the use of tax abatements, however, the city's commitment of resources to downtown redevelopment far outweighed its neighborhood revitalization efforts (Keating et al., 1989; Keating et al., 1995). This policy reflects the long-standing influence of business on city politics.

Notwithstanding the extensive use of federal programs like Model Cities, CDBG, low-income housing tax credits, job training, and others aimed at the problems of poor urban neighborhoods and their residents, Cleveland's poverty rate grew from 17% in 1970 to 30% by 1990. By the mid-1990s, more than 40% of Cleveland's population was poor, with poverty spreading to more of its neighborhoods and long-term poverty increasing (Coulton & Chow, 1995). Cleveland's public housing became housing of last resort to its mostly black population (Chandler, 1995). A majority of the shrinking, predominantly black student body in the public schools was below the poverty line. Black male unemployment consistently exceeded national rates.

All these dismal data were indicators of what has been termed *uneven development* between the downtown and urban neighborhoods of central cities (Keating et al., 1989, 1995; Swanstrom, 1985). The result is what has been termed the *dual city,* in which downtown areas can thrive, and the white-collar sector mostly employing white suburban commuters can grow, whereas poor neighborhoods and their working class and poor residents, especially blacks and Hispanics, do not share in this growth and must cope with high rates of poverty, unemployment, and other social problems (Goldsmith & Blakely, 1992).

Unlike Cleveland in its industrial heyday, there are few signs that downtown redevelopment, the beneficiary of massive public subsidies, has generated enough jobs for low-income and unskilled city residents. This raises the issue of the costs and benefits of this strategy and the politics that support it.

## Downtown Redevelopment Politics in Cleveland

Redevelopment politics in Cleveland represent the long-standing efforts of the civic elite to promote prosperity through public financing of private business. In arguing for the support of the city's elected officials and their constituents for such policies, the benefit has always been cast as the retention and expansion of jobs, primarily if not exclusively for the city's own residents.

As Bartimole (1995) recounts, however, this has all too often resulted in great benefits to business without the accompanying promised public benefits. He describes the influence of organized private interests over the city's elected officials in keeping with urban regime theory. This posits the considerable influence of private interests, through lobbying, campaign contributions, and threats to relocate business, over local public officials, who have usually been willing allies (see, e.g., Elkin, 1987). Bartimole describes how the downtown business interests (major corporations, utilities, banks, law firms, real estate developers, parking owners, etc.), civic leaders, and politicians promoted the city's urban renewal program and its successors (Bartimole, 1995).

Never was this influence more dramatic than in the defeat of populist mayor Dennis Kucinich in 1979 after a turbulent 2-year term. Refusing to bow to business demands for tax abatements for downtown redevelopment and the sale of the decrepit municipal public power system to its private competitor, Cleveland Electric Illuminating (CEI), Kucinich saw the banks refuse to refinance the city's debt, resulting in the city's default in 1978. Kucinich barely survived a recall attempt, only to be badly defeated by Republican George Voinovich in 1979. Voinovich was the choice of the business and civic establishment unified to eliminate Kucinich as an obstacle to the continuation of public support for private redevelopment projects (Bartimole, 1995; Swanstrom, 1995). Magnet (1995) described this as a benign conspiracy of a self-described cabal of corporate executives led by the retired CEO of the Eaton Corporation to overthrow the Kucinich administration. This was formalized in 1980 as "Cleveland Tomorrow," a council of the chief executive officers (CEOs) of the region's largest corporations (Shatten, 1995). Cleveland Tomorrow has played a key role in virtually all major policy decisions regarding redevelopment strategy for Cleveland since 1980. Kucinich's legacy was the preservation of the municipal public power system, even though it had to weather later attacks by CEI (Keating et al., 1995).

Business interests have strongly supported Voinovich and his successor as mayor, Michael White. Their primary interest has been in promoting economic redevelopment and growth, focused on large-scale projects in downtown Cleveland. A succession of power brokers led by corporate lawyers and CEOs has represented business interests. They have advocated governmental tax subsidies such as tax abatements, transit, and infrastructure improvements to promote new development and private investment. Their commitment to corporate investment in the public schools and urban neighborhoods has paled in comparison.

Cleveland's business community has always had tight-knit leadership. Working through civic institutions like the Union Club and philanthropic foundations, the business community has played a key role in decision making, often out of the public eye. This has been true especially when public support has been needed for major projects. Although generally unified, the business community has sometimes divided over development priorities and the timing, financing, and location of certain projects.

# The Politics of Sports
# Stadiums in Cleveland: Gateway

## Introduction

American cities have redevelopment strategies that have emphasized the downtown as the focus for office, retail, and entertainment activities (Frieden & Sagalyn, 1989). Those cities pursuing a convention and tourism strategy have emphasized tourist attractions, including sports. Combined with a growing trend toward moving professional sports franchises, this has resulted in a bidding war among cities competing to retain or attract these franchises (Euchner, 1993; Shropshire, 1995).

Central to this competition has been the demand by team owners for new facilities and very favorable leases (especially control of revenue from luxury loges and concessions). Chicago, for example, has had to deal with demands from five different professional sports franchises (Euchner, 1993; Mier, 1993). With a few notable exceptions, these new sports stadiums are city owned or county owned and involve substantial subsidies, both for construction and in the lease arrangements with team owners. Perhaps the most extreme recent examples were the construction of new stadiums in Indianapolis, St. Louis, and St. Petersburg, Florida, just to try to win franchises (Zimbalist, 1992). Indianapolis wooed the football Colts from Baltimore and, more recently, St. Louis has become the home to the Los Angeles Rams (having previously lost the football Cardinals franchise to Phoenix). Even those teams that have been the beneficiaries of such public largesse, despite signing long-term stadium leases, have still left open the possibility of later threatening to relocate elsewhere (depending on such factors as attendance and revenue guarantees).

This intercity competition (very similar to the competition for other types of economic development involving public subsidies) is premised on the image of cities striving to be among the top or second tier of major cities to promote growth and the supposed economic benefits of successful sports franchises. With the expansion of the major professional sports leagues, more franchises have become available, allowing more cities to have major league professional baseball, basketball, football, and ice hockey franchises. Nevertheless, losing cities like St. Petersburg, which failed in its effort to lure the White Sox from Chicago (Euchner, 1993), continue to woo existing franchises. Oakland, which lost the football Raiders to Los Angeles a decade ago, won them back in 1995 (with promises of major revenue increases from luxury box seats in a remodeled county-owned stadium shared with the baseball Athletics, a team that Oakland lured away from Kansas City). The political and business leadership of those cities threatened with the loss of a team (for example, San Francisco and the baseball Giants) have gone to great lengths to avoid this fate for fear that this would hurt their city's image (DeLeon, 1992). Several of those cities that lost professional franchises have then replaced them with expansion teams (e.g., Kansas City, Milwaukee, New York, and Seattle in baseball).

## Cleveland: Downtown Redevelopment and Sports

In Cleveland, this debate has been prominent for more than decade, as the city lost its ice hockey franchise and faced the threat of losing the baseball team (Indians) and, more recently, the football team (Browns).

The Browns announced their planned departure for Baltimore in November 1995, on the very eve of passage of a tax referendum that would extend an existing stadium sin tax to renovate the city's municipal football stadium. Mayor White, with strong support from the media and business and civic leaders, led a crusade to persuade the National Football League (NFL) to prevent the departure of the Browns or to replace the franchise. Baltimore offered the Browns a rent-free new stadium (estimated to cost $200 million) and training facility, revenue from the concessions, including luxury loges, and generous moving expenses. Cleveland sued both the Browns and the Maryland Stadium Authority.

In February 1996, the NFL and the city of Cleveland settled this dispute. In return for the city dropping its lawsuits, the city retained the team name and was guaranteed a replacement franchise no later than 1999, with the NFL providing an unprecedented construction loan for a new football stadium (estimated to cost between $220 million and $250 million) to cover in part the cost difference between a new versus renovated facility (Koff, Heider, & Grossi, 1996; Sandomir,

1996). The city had already raised $280 million from the sin tax extension, with increased downtown parking taxes and other tax increases estimated to raise an additional $61 million, to provide public financing for a modern football stadium.

Providing state-of-the-art stadiums for Cleveland's three major professional teams and how to pay for them have been pivotal issues in downtown development. This case study will examine the politics of the Gateway stadium/arena project.

## Cleveland's Sports Franchises—Indians and Cavaliers

The Cleveland Indians baseball club traces its origins to the Cleveland Spiders (1887-1899). With the formation of the new American League in 1901, Cleveland rejoined the major leagues with a team eventually renamed the Indians. In 1931, Cleveland's Municipal Stadium was completed. The Indians moved to the stadium permanently in 1947. The team went to the World Series in 1920, 1948, and 1954. After prospering from 1948 to 1959 on the field and at the gate, the Indians fell on hard times for most of the following three and a half decades, until their renaissance in 1994-1995. After making it to the World Series again in 1995, the Indians soon sold out the entire 1996 season.

In 1970, Cleveland gained a National Basketball Association (NBA) expansion franchise named the Cavaliers. In 1974, owner Nick Mileti moved the team from the Cleveland Arena downtown to a new Coliseum, located halfway between Cleveland and Akron. As the NBA increased in popularity, the Cavaliers enjoyed success under the ownership of the wealthy Gund brothers, who purchased the team from Ted Stepien in 1983, after the team became nearly bankrupt (Grabowski, 1992).

## Domed Stadium

In 1983, the fortunes of the Cleveland Indians were at a low ebb and its major owner (Steve O'Neill) died, leaving the future ownership of the franchise in doubt. Without any immediate overt threat of losing either of the teams occupying Cleveland's municipal stadium, an ambitious Republican Cuyahoga County Commissioner proposed building a domed stadium to replace the city's aging Depression-era stadium, to be financed solely by public funds, namely, a special countywide property tax.

Attacked as an overly expensive and regressive tax, the proposal was soundly defeated in a 1984 referendum. This result can be attributed to a variety of factors: public skepticism about the need for the facility (including the dome, which added considerably to the estimated cost), the absence of a credible threat of the loss of either the baseball or football team, a division among civic and political leaders over the proposal, and, probably most important, voter opposition to additional property taxes.

## Gateway

In 1986, the Cleveland Indians were sold by the O'Neill estate to David and Richard Jacobs on condition that they remain in Cleveland (at least 5 years). At that point, attendance was at a low ebb, reflecting the team's miserable record (Grabowski, 1992). The Jacobs brothers were wealthy real estate developers based in suburban Cleveland. (In 1995, Richard Jacobs's wealth was estimated at $380 million, making him one of the 400 wealthiest individuals in America. David had died in 1994; Forbes, 1995.) Their company had developed the Galleria, a federally subsidized downtown shopping mall, the first of its kind for Cleveland, and in 1988 they announced plans for building two downtown office towers, together with hotels. The Jacobs brothers were hailed as heroes by the Cleveland media for their investments in Cleveland. Amidst controversy, they demanded and received full, 20-year property tax abatements (worth an estimated $225 million) for their office-hotel project, together with other public subsidies (Keating et al., 1989; Keating et al., 1995). Although the Society bank office tower and Marriott hotel were built and opened in 1992, the proposed Ameritrust building was postponed because of a downturn in the real estate market (and the merger of Ameritrust Bank with Society Bank).

The major loser in this arrangement was the Cleveland public school district, heavily dependent on property taxes. It has long struggled with inadequate finances, a huge dropout rate and low student performance, reflecting the poverty of a large number of its students. The district has also been buffeted by a school desegregation order in effect since 1977, steadily declining enrollment (from 132,000 in 1973 when the desegregation case began to 74,000 in 1995), administrative scandals, a regular turnover of superintendents, and political squabbling. The last time that Cleveland voters approved a property tax operating levy in support of their public schools was 1983 (and before that, 1970). In 1994, Cleveland voters twice rejected school operating levies.

In early 1995, the incumbent school superintendent resigned abruptly, and in March a federal district court judge, instead of lifting the desegregation order (including mandated busing) as requested by the school board, placed the deficit-ridden district under state receivership. The tarnished reputation of the public schools was second only to crime as a deterrent to owning a home in the City of Cleveland in surveys of home sellers and home buyers (Bier, 1995). The contrast between the sorry plight of the impoverished public schools compared with the massive expenditures for downtown development would be raised periodically during the debate over the need to subsidize sports stadiums for the wealthy owners of Cleveland's professional franchises.

In November 1989, Michael White, an upstart state representative and former Cleveland City Council member, defeated Cleveland's powerful City Council President George Forbes in the race to succeed George Voinovich (now governor of Ohio) as mayor. White campaigned on the need to improve the city's neighborhoods and public schools. After his upset victory, he appointed a strong neighborhood advocate to run the city's community development programs and he backed a reform slate of school board candidates, who were victorious in 1992.

Almost immediately, however, he turned his attention to the proposed Gateway project. After they purchased the Cleveland Indians in 1986, the Jacobs brothers had pressured the city to build a new, baseball-only stadium to allow the team to move out of the city-owned, multipurpose stadium. Their motivation was undoubtedly partly to escape the management of that stadium, controlled by Art Modell, owner of the football Browns, and to be in a facility that would provide them with revenue from luxury loges and concessions. Implicit was the threat that they would sell the team to a new owner who would move it to another city on the expiration of their lease. This threat was made explicit shortly before the vote on taxes to finance a new stadium when then-baseball commissioner Fay Vincent appeared before the Cleveland City Council to warn of this consequence should the voters defeat a stadium tax proposal.

Modell, the owner of the Browns, did not join in the Jacobses' request for a new facility. His team was enjoying success on the field and at the box office, filling most of the cavernous stadium, which holds almost 80,000, during the 1980s. By 1994, however, after Gateway's opening, an envious Modell demanded a new or renovated football stadium for his franchise.

The basketball Cavaliers were wooed by the city with the promise of a new arena. They continued to occupy the Richfield Coliseum, owned by the Gund brothers. George and Gordon Gund are the wealthy sons of George Gund, the late president of the Cleveland Trust Bank (later Ameritrust and since merged with Society Bank). Their estimated combined wealth in 1995 (with a brother and sister) was $1.1 billion, making them one of America's richest families (Forbes, 1995). They owned Cleveland's short-lived National Hockey League franchise in the 1970s (Grabowski, 1992). It was argued that bringing the basketball franchise back to downtown Cleveland from its suburban location would further enhance downtown revitalization. Because the Gunds owned their own facility already and it did not need replacement, they had to be enticed to return their successful franchise to Cleveland.

The stage was set for what was called Gateway—a baseball-only stadium for the Indians seating 42,000 and a new adjacent arena seating 21,000 for the basketball Cavaliers to be built on the site of the central market, which was to be relocated by the city (Hirzel, 1993). Mayor White became the most visible political advocate for the project, supported by the Cleveland City Council and the three Cuyahoga County Commissioners.

Gateway was to be financed by a 15-year special county "sin tax" or excise on alcoholic beverages and cigarettes, estimated to bring in $275 million. This eliminated potential opposition to countywide blanket increases in either property or sales taxes. Ohio's liberal Democrat governor, Richard Celeste (a Clevelander), who opposed the proposal to build a domed stadium in 1984, also promised state assistance. The estimated construction cost of the new facilities was claimed to be approximately $280 million. At no point did the Jacobs or Gund brothers publicly agree to contribute to the financing of the facilities for their teams, except for modest lease payments. Both would later pay for the naming rights for their facilities—now known as Jacobs Field and Gund Arena. The major private "contribution" came in the form of a loan from Cleveland Tomorrow. Rather, the public sector was primarily left to bear the burden of construction costs for the stadium, arena, administration building, and garages.

Business and civic leaders and the media, led by the city's only major daily newspaper (the *Plain Dealer*) rallied behind Gateway. Proponents quickly raised $1 million to support the tax referendum, much of that money coming from the baseball team (approximately $200,000) and the basketball team ($100,000) and the rest from corporate supporters. Voters were promised "28,000 good-paying jobs for the jobless; neighborhood housing for the homeless; $15 million a year for schools for our children; revenues for city and county clinics and hospitals for our children; revenues for city and county clinics and hospitals for the sick; energy assistance programs for the elderly." Gateway would not seek a tax abatement but rather would pay property taxes (Bartimole, 1994).

In the face of these claims for economic development benefits of the project and keeping the Indians and bringing the Cavaliers back to downtown Cleveland, there was a poorly financed opposition led by the United Auto Workers (UAW) local. The UAW had previously opposed tax abatements for downtown development (including the Jacobs brothers' projects), although unsuccessfully. Opponents pointed to the need for the use of any new taxes for more pressing social service needs, especially in light of the city's precarious fiscal situation and cutbacks in urban aid by the state and federal governments, and to support the public schools.

In the May 1990 referendum, the Gateway sin tax passed but only by 51.7% of the 383,000 votes cast. Although suburban voters generally favored the sin tax (by a margin of 55% to 45%), voters in all but one of the city of Cleveland's 21 wards voted against the sin tax and Gateway (and the position of their popular Mayor) by an overall margin of 56% to 44%. The support of most of the civic, business, and political establishment of greater Cleveland (with the notable exception of powerful black Congressman Louis Stokes) confirms urban regime theory. Powerful private interests, supported by elected local officials for whom they provide financial support, virtually dictate the municipal agenda, in this case one centerpiece of downtown redevelopment. The opposition could not overcome the prodevelopment arguments, despite the refusal of a majority of city voters to support the tax. In 1993, White easily won reelection, virtually unopposed but with a campaign war chest of more than $1 million, largely provided by Cleveland business interests (Bartimole, 1994).

Once the sin tax was passed, the Gateway Economic Development Corporation (GEDC), a nonprofit, was formed to oversee construction. Although the city and county were critical to the financing of the project and were represented on GEDC's board, this quasi-public entity operated largely out of the public limelight, with minimal accountability and oversight. Corporate attorney Tom Chema, an ally of Ohio Governor Celeste, was put in charge. James Biggar, former chair of Nestle USA, chaired the GEDC board.

Pressure was intense to complete construction in time to open the facilities in 1994. The result was predictable. The original construction estimates rose. County-guaranteed construction loans of $120 million became critical to the project's financing. The City of Cleveland piggybacked construction of a Gateway garage onto a new garage at City Hall in addition to infrastructure improvements. Instead of a 50% to 50% public-private split in financing construction, as originally promised to the voters, the public share could eventually be as high as 70% (Kissling, 1992). To make Gateway's financial projections workable, the city and county had to have the Ohio Legislature grant Gateway property tax abatement, thereby undercutting one major claim of Gateway's prospective economic benefits to the city by its proponents in the 1990 referendum. Bartimole caustically noted,

> The high-powered maneuvering of Cleveland's cohesive downtown estab-
> lishment and its puppet-like city government to accommodate, with hundreds
> of millions of public dollars, these two sports—each of which is owned by
> multimillionaire brothers on the Fortune list of the wealthiest Americans.
> (Bartimole, 1994, p. 30)

In 1994, the Cleveland Cavaliers franchise was estimated to be worth $133 million, making it the sixth most valuable NBA franchise, compared with an average franchise value of $111 million (Ozanian, 1995). The Cleveland Indians franchise was estimated to be worth $103 million, making it the thirteenth most valuable baseball franchise, compared with an average franchise value of $111 million (Ozanian, 1995). With their occupancy of their new facilities with very favorable leases and high attendance, their value has undoubtedly increased.

Although the *Plain Dealer* mostly applauded Gateway's rapid progress, Roldo Bartimole, Cleveland's noted muckraker (Sweeney, 1994), was a lonely voice in exposing cost overruns, decrying lavish spending for special features for the teams, and criticizing the lack of public scrutiny.

Gateway opened on schedule in April 1994 to the applause of the citizenry and Gateway proponents and an opening day appearance by President Clinton. The Cleveland Indians enjoyed their best season in decades with tremendous attendance. Unforeseen trouble soon arose, however. In midsummer, baseball players went on strike, closing Jacobs Field down. Then, in October 1994, the *Plain Dealer* exposed a financial scandal involving the Cuyahoga County Treasurer's bond investment fund (called SAFE). This resulted in the liquidation of SAFE and estimated losses of $115 million to the County and other participating local governments. Nevertheless, the County was required to make payments on Gateway's County-guaranteed construction bonds. The County subsequently had to cut its budget by 11% for 1995 to 1998 to cover its SAFE losses.

Although the Gund Arena also opened on schedule in that fall and baseball resumed after the end of the strike in May 1995, Gateway's troubles continued. Cost overruns finally became headline news in the *Plain Dealer* in Spring 1995. The final combined construction costs of Gateway are estimated to be approximately $464 million. This does not count interest payments on the construction bonds or such related public investments as a city-constructed garage, a Regional Transit Authority (RTA) financed tunnel connecting the nearby Tower City project (also heavily subsidized), and public infrastructure improvements.

In June 1995, as cost overruns of $22 million for Gund Arena and $6 million for Jacobs Field were revealed and the GEDC could not pay these bills, Chema resigned. His successor publicly demanded that the teams pay these debts (Rutchick & Koff, 1995). The County, citing its SAFE losses, declined to bail out GEDC. The *Plain Dealer* revealed the additional construction costs attributable to team demands under the very favorable contract that they negotiated before they would sign leases (Rutchick, Heider, & Koff, 1995). These

revelations had long been reported by Bartimole but his accusations and warn-
ings of cost overruns were ignored.

Even if the two teams contribute, it is likely that the public will pay the
majority of the $22 million. As 1995 ended, Cuyahoga County agreed to make
an interest-free $11.5 million loan and guarantee a private loan to cover the
remainder (Heider, 1995). The city's convention and visitors bureau and its
business supporters balked, however, at an effort to divert $550,000 annually
for 10 years from its budget to help to finance this elimination of Gateway's debt
(Koff, 1996). In February 1996, the two teams were forced to contribute to the
operating costs of Gateway, beyond their lease payments, because of GEDC's
insolvency (Gillispie, 1996).

What are the lessons of Gateway? First, although public bodies (the city,
county, and state) will most likely end up paying the majority of Gateway's
construction costs, its design and amenities were virtually dictated by its private
tenants. The pressure of private interests (the powerful and wealthy franchise
owners and their allies in the business community and media) greatly influenced
the decisions of elected officials. Second, the "local state" (in this case, the city
of Cleveland and Cuyahoga County, aided by the State of Ohio) is essential to
the success of "the city as growth machine" (Logan & Molotch, 1987). Local
governments were essential in providing construction subsidies and loan guarantees, acquiring the site, relocating the commercial tenants, and providing
necessary infrastructure. All this was done on terms highly favorable to wealthy
private interests that risked the least.

Finally, the 1990 referendum campaign promises of additional tax revenues
and thousands of new permanent jobs ring hollow. Tax abatement for Gateway
foreclosed the former, except for the possibility of increased taxes from redevelopment in the commercial district adjacent to Gateway. Likewise, this is
where jobs would have to be generated. As of early 1996, although some
redevelopment has occurred, there is no evidence yet of substantial economic
benefits to the city and county.

Yet, the need, promoted by the civic establishment, for sports franchises to
promote Cleveland's image and, therefore, to be subsidized, remains paramount.
Ironically, Cleveland's national and internationally acclaimed symphony may
enhance the city's image more than its sports franchises. The orchestra is
currently embarked on a fund-raising campaign to modernize its symphony hall
at a cost of $20 million. Meanwhile, festering social problems that negatively
impact on the city's economic well-being remain unresolved. In the face of
threatened federal budgetary cutbacks in social programs, the future prospects
for addressing these issues are not promising. Cities like Cleveland face harsh
choices in public spending—public schools, crime prevention, job training,

infrastructure repairs, and neighborhood revitalization must compete with tax abatements for business and subsidies for blockbuster public and private projects to enhance downtown redevelopment.

## Conclusion

The Cleveland saga of wealthy private owners of professional sports franchises using their influence, that of their civic and business allies, and the threat of leaving the city unless the local governments finance new facilities on their terms and timetable is hardly new or very different from the experience of most other cities. In 1995 alone, baseball franchises demanded and won promises of public financing of new stadiums in Detroit, Milwaukee, and Seattle. In Seattle, the city and county promised a new retractable roof stadium (at an estimated cost of $325 million) in the face of threats to move the franchise after its most successful season, despite voter rejection (although by only one fifth of 1%) in September 1995 of a sales tax increase to finance the public share of the costs. Then, in early 1996, the football team, the Seattle Seahawks, announced their intention to break their lease in the city's Kingdome (beset by physical dilapidation) to relocate to Los Angeles. Houston's baseball franchise threatened a future move unless its demands were met, even as the Houston football franchise was on the verge of deserting the city for Nashville, Tennessee. Public financing for new baseball and football stadiums in Cincinnati was approved in a March 1996 referendum.

Promoting sports as beneficial economic development for the city, region, and its citizens to justify massive public expenditures, without which most of these sports stadia would not be built, has become routine. The media have accepted this argument as dogma and parroted the owner's threats of the consequences of a city's failure to meet their demands.

San Francisco, where voters twice narrowly rejected referenda in 1987 and 1989 to finance a new baseball-only stadium for the Giants, is one such exception (DeLeon, 1992). DeLeon notes that former liberal mayor Art Agnos, subsequently defeated in his reelection bid in 1991,

> damaged his reputation as a progressive mayor by rushing to placate millionaire [Giants owner] Bob Lurie, by currying favor with downtown business elites, by spouting progrowth rhetoric in selling a new ballpark to the voters, by angrily and vindictively attacking those who had defeated his proposal, and by abandoning environmentalists, neighborhood preservationists, and housing activists on their small beachhead of social change in China Basin. (DeLeon, 1992, p. 123)

The Giants remained in San Francisco under new local ownership. In 1995, the Giants unveiled a new plan to finance a stadium costing $255 million without resorting to direct public subsidies (Epstein, 1995). A referendum passed in March 1996.

San Francisco is a rare example of a successful grassroots political challenge to the basic tenets of the city growth machine through the sports model. In contrast, citizen opposition to Baltimore's new baseball stadium failed badly (Euchner, 1993). San Francisco and Seattle are the only U.S. cities to limit downtown office development, in both cases by voter initiatives in the face of urban regimes arguing for downtown growth (Keating & Krumholz, 1991). In most cases, the citizenry either silently witnesses the expenditure of public subsidies for private sporting interests or willingly agrees to tax itself, typically with little promised in direct, guaranteed benefits to the public bodies involved in stadium financing.

The case of Cleveland and the Gateway project illustrates the local state's key role in financing public-private partnerships promoted by urban regimes. Local governments, even in progressive political climates like San Francisco and Seattle, have been driven to pursue private sports franchises at great public expense to compete with other cities under similar pressure from civic elites. With little evidence that the citizenry, particularly poor residents, benefit significantly from increased employment or improved public services attributable to revenue from sports stadia, projects like Gateway are monuments to intercity competition for prestigious sports franchises supposedly beneficial to economic development, which primarily benefit the wealthy owners. Sometimes this pits competing local economic interests against each other. In Cleveland, for example, the bottling industry opposed the Gateway excise tax on alcohol.

Ironically, the configuration of these new stadiums and increased prices may limit the ability of many city residents to even attend games. St. Louis pioneered the sale of personal seat licenses (the right to purchase season tickets) to help finance its new domed football stadium. Sale of these licenses is likely to be part of the financing scheme for Cleveland's new football stadium (Vickers, 1996). Overall, instead of benefiting the entire city, publicly subsidized sports stadiums contribute to the American city as the dual city of advantaged and disadvantaged citizens.

What does this case study tell us about the relationship between regime theory and regulation theory? It reinforces the pattern of cities competing in a global economy but facing the threat of losing footloose businesses seeking higher profits where competing cities offer enticing subsidies. Indeed, U.S. professional sports leagues have taken on something of an international look. Baseball has expanded to Canada, with the possibility of adding teams from Japan and

Mexico. Professional basketball has also just expanded to Canada. The NFL sponsored a World Football League, based in the United States and western Europe.

Place-bound U.S. cities seeking economic growth are pushed by urban regimes to provide desirable magnet amenities, both public and private, with public subsidies to promote private investment. Despite what often are massive public subsidies resulting in private gain and little discernible public benefits, there are few examples of successful citizen-based opposition. Sports stadiums in Cleveland and elsewhere in U.S. cities are but one example of this phenomenon.

C
H
A
P
T
E
R 11

Regulating Suburban Politics

*"Suburban-Defense Transition,"*
*Institutional Capacities, and*
*Territorial Reorganization in*
*Southern California*

ANDREW E. G. JONAS

There can been little doubt that urban regime theory and regulation theory have reinvigorated studies of governance in capitalist states. Yet despite a common interest in governance, only recently have attempts been made to achieve theoretical commensurability between the two approaches (Painter, this volume). This is perhaps because regulation theory and urban regime theory have hitherto dealt with two different spatial scales, national and urban, respec-

AUTHOR'S NOTE: The financial support provided by the National Science Foundation (Award No. SBR-9512033) and a Chancellor's Summer Research and Teaching Award from the University of California at Riverside is gratefully acknowledged. Thanks to Catherine Lee for research assistance, to Tom Feldman, and Amer Althubaity for intellectual support, and to Bob Beauregard, Cindy Horan, Mark Goodwin, and Mickey Lauria for constructive criticisms of an earlier version of the manuscript. Richard Florida originally made me aware of the "suburban-defense economy" and some of his ideas are reflected in this chapter. The usual disclaimers apply.

tively. This state of affairs seems to be changing, however. On the one hand, urban regime theory has broken through the theoretical impasse reached by community power structure analysis by situating the "new urban politics" in its national and global contexts, albeit the theory remains curiously silent about the nature of the structures of accumulation into which urban localities and governing coalitions are inserted (Cox, 1993). On the other hand, regulation theory has paid more attention to these wider structures, arguing that recent transformations in local governance reflect fundamental changes under way in modes of economic and social regulation in capitalism. So far, however, regulation theory has failed to provide much more than a structural context for recent transformations in local governance. Jessop, who has advocated a strategic-theoretical approach to the study of governance and economic regulation, puts it at follows:

> It is hardly surprising that there are, as yet, no adequate regulationist explanations of the structural transformation and/or strategic reorientation of the local state. At best we have more or less plausible regulationist contextualizations of these shifts. Yet, however detailed the analysis of the strategic context might be, it cannot itself generate an adequate explanation for strategic action. This would require in addition at least some account of the strategic capacities of actors (individual and/or collective) to respond to economic problems, the strategies that they try to pursue and the relationship between these capacities and the strategies and those of other relevant actors in that context. (Jessop, 1995b, p. 321)

Jessop does not elaborate on how one might go about investigating the strategic capacities and strategies of local actors. The possibility that urban regime theory might fill this gap in regulation analysis is encouraging efforts to achieve theoretical commensurability between the two approaches.

My argument in this chapter is that, if there ever is to be a productive engagement between regulation and urban regime theories, the relationships between structural context and strategic action must be interrogated through the analysis of actual transformations in urban governance in particular spatial settings. One approach, then, would be to focus on a locality or region which occupies a strategic position in a national economy and is in a period of crisis and restructuring. Within this context, it should be possible to investigate transformations under way in the local state, focusing on local actors, their strategic capacities and strategies, and struggles. An important component of the investigation would be to actuate and contextualize the regulationist concept mode of social regulation in a way that draws out the relationships between local capacities and strategies, and the wider structures of accumulation and regulation into which local actors and interests are inserted. It should then be possible to focus explicitly on the interconnections between the national and the local, rather than treat these as separate scales of analysis.

Conceptualizing the role of spatial scale in recent local political and economic transformations has been problematical for regulation theorists and all but ignored by urban regime theorists. Although regulationist theorists have focused increasing attention on the contribution of regional governance to industrial transition (Scott & Paul, 1990), they continue to ignore corresponding mechanisms of social regulation at the regional scale. With a few exceptions (Saxenian, 1994), regulationist interpretations of regional governance are economistic, ignoring the interests, conflicts, and strategies involved in the coalition-building process. Meanwhile, urban regime theory has confined its focus to urban governance and has made little contribution to understanding processes of coalition building and territorial reorganizations at wider spatial scales, even though these clearly can have effects on urban policies (Jonas, in press). In general, neither regulation nor regime theory has done a particularly good job of drawing out the relationships between context, strategy, and scale in transformations of the local state.

To the extent that regulatory transformations do play out at different spatial scales, then territorial reorganizations within the state need to be brought to the fore in strategic-theoretical analysis. My intention in this chapter is to explore the relationships between context, strategy, and scale in the transformation of local government in Southern California, or more precisely the Inland Empire region, which comprises 26 or so cities and the unincorporated areas of western Riverside and San Bernardino counties. The San Bernardino-Riverside standard metropolitan statistical area (SMSA) (1990 population 2.59 million) is in the midst of a period of crisis, restructuring, and political transformation. The nature of the crisis and transformations underway reflect the region's strategic but changing position in the postwar "suburban-defense economy" (Florida & Jonas, 1991). In particular, problems of local government fragmentation and interjurisdictional competition have made the issue of territorial reorganization salient to the regulatory problematic now facing the region, and has led to an increasing emphasis on building coalitions and institutions of governance at the regional scale.

The first section of the chapter reviews what regulation theory and urban regime theory have contributed to an understanding of the role of spatial scale in political and economic transformations. I then establish the structural context for the present case study. This involves contextualizing the crisis as a "destructuring" of the suburban-defense economy that formed the dominant mode of social regulation in U.S. fordism. The salient elements of this mode of social regulation are identified and connections are established between the national and local contexts. The chapter then focuses on the politics underlying the "suburban-defense transition" in the Inland Empire region of Southern California. This involves investigating the strategic capacities and strategies of local actors and institutions. Two sets of capacities and strategies are discussed: (a) those governing the suburban land development process and (b) those regulating local economic development and patterns of inward investment. In both sets of cases, I emphasize the relationships between context, strategy, and scale by looking at the ways in which regional approaches rather than specifically local strategies have been pursued by local actors. The chapter concludes with some general observations about the prospects for integrating the regulationist and urban regime schools in the study of urban politics, and political transformations at wider spatial scales.

# Regulation Theory, Urban Regime Theory, and Spatial Scale

## Regulation Theory

Regulation theory has developed as way of explaining why crisis tendencies endemic to capitalism do not necessarily always materialize in particular national settings (Aglietta, 1979; Lipietz, 1987). The theory suggests that such crisis tendencies are held back by the development of national institutional frameworks that bring about a structural coherence to the dominant "regime of accumulation" (RA). These institutional frameworks—"modes of social regulation" (MSRs)—regulate the struggle between labor and capital around the production and distribution of surplus value. When fully integrated with its corresponding RA, an MSR realizes a relatively stable allocation of social product between production, investment, and consumption.

In strategic-theoretical accounts of regulation, the form of an MSR is seen to be shaped by historically and geographically contingent circumstances that vary within and between different national contexts (Jessop, 1990b; Tickell & Peck, 1995). In this respect, an MSR both regulates, and is the outcome of, nationally specific accumulation strategies and political struggles. Political struggles are contingently related to dominant accumulation strategies (Clarke, 1988), so it is unlikely that an MSR is ever fully integrated with its corresponding RA. Rather it will be possible to find a variety of potentially contradictory regulatory and counterregulatory (Painter, this volume) processes operating within a national economy that tend to bring about a coherence in the allocation of social product between production, investment, and consumption, or conversely tend to undermine the structural coherence of the MSR.

Uneven development across different spatial scales is a central part of social regulation in capitalist states, in terms of both the variations found in MSRs between national economies as well as within them. Recent regulation scholarship has tended to emphasize international differences in MSRs, and in particular national variations from classic fordism in the United States (Peck & Tickell, 1992; Tickell & Peck, 1995). This international emphasis is quite understandable given the ravages of economic globalization and neoliberalism, and the implications of international regulatory frameworks for the process of hollowing out in the nation-state during the current crisis of fordism. Thus, there has been a great deal of speculation about the nature of the RA that will succeed, or already has replaced, fordism. Quite apart from the issue of whether or not fordism is in fact in crisis (see Lovering, 1990), speculation has so far generated little debate about what a postfordist MSR might look like or at what spatial scale(s) it should be organized (Tickell & Peck, 1995). Suffice it to say, the international and national scales will continue to play strategic roles in shaping regulatory and counterregulatory frameworks in postfordism.

By comparison to the interest in the national and international scales, regulation theory has had relatively little to say about uneven development at the subnational scale (Peck & Tickell, 1992; Painter & Goodwin, 1995). Peck and Tickell (1992) argue that the establishment of "local modes of social regulation" may be necessary to regulating uneven development within states, and constitutes one type of spatial fix to capitalism's crises. To that extent, regulationists recognize scale as a force of social regulation in capitalism, albeit there has been little evidence provided of the politics of scale in regulatory transitions.

Tickell and Peck (1995) are currently skeptical that it is possible to identify coherent "local modes of social regulation" emerging from the crisis of fordism. Neoliberalism champions market forces over reregulation, and emphasizes interurban competition over the forging of progressive local social contracts between labor and capital. Cities and local business elites are consequently engaged in a beggar-thy-neighbor struggle for inward investment and jobs. This struggle tends to accentuate patterns and processes of uneven development, making it difficult to realize a coherent spatial fix at the local scale.

There are then important connections between the wider (de)regulatory context and the form of politics at the urban scale. The crisis of fordism is associated with widespread transformations in local government, and an increasing emphasis is placed on formal and informal partnerships between local public sector and private sector actors as means of regulating urban development in the 1990s (Cochrane, 1991; Goodwin, Duncan, & Halford, 1993; Painter, 1991).

Regulationist interpretations of these transformations have, however, tended to portray them in rather stark and simplistic terms, ignoring the strategic capacities and strategies of local actors. Urban regime theory can provide a richer picture of the local interests, struggles, and strategies involved in urban political transformations, although it too is silent on certain issues of strategic significance, including territorial reorganizations within the state.

## Urban Regime Theory

Urban regime theory is mainly interested in finding answers to the question "who governs cities?" (Stone, 1989, 1993). But unlike an earlier generation of city power structure studies (Dahl, 1961; Hunter, 1953), regime theory's explanation for the "capacity to govern" is not limited to identifying the groups that win local elections (pluralism) or those that control resources (elitism). Rather urban policy outcomes result from a more complex set of interactions /conflicts that determine (a) which groups/actors are able to form a governing coalition, (b) what sorts of relationships are developed among coalition members, and (c) what kinds of resources can be deployed by members of the governing coalition to shape public policy (Stone, 1993, p. 2).

Regime analysis bases its theory of governing coalitions and policy formation on the structural need for private interests to overcome the division of labor between state and market (Elkin, 1987). Thus, local government faces an enduring dilemma between the need for tax revenue accumulation and demands placed on public expenditures by the electorate (Calaveta, 1992; Piven & Friedland, 1984). City governments tend to embrace the development agenda of downtown business interests even though revenue-generation activities may bring elected officials into conflict with community groups (Stone, 1989). One reason for this is the increasing penetration of global market forces into urban areas that structurally predispose actors dependent on the local economy to pursue a proeconomic development policy agenda (see, however, Horan, 1991).

The overwhelming empirical focus of regime theory is on downtown business-dominated governing coalitions that can be more or less socially inclusive and democratic. We also learn, however, that the types of businesses that tend to dominate such coalitions are those "whose revenues depend on the level of economic activity in the city *and* [italics added] metropolitan region" (Elkin, 1987, p. 90). This subtle shift in scale emphasis is not problematic insofar as the types of businesses based in downtowns—banks, insurance companies, commercial brokers, real estate, newspapers, utilities, and so forth—have metropolitan-wide economic interests. But it becomes more problematic when one considers that the territorial scale and scope of business economic interests can influence the coalition-building process, and hence also types of policies pursued by the coalition. In other words, regime theory is extremely ambiguous about the relationship between the territorial scope and scale of private business interests and policy choice.

Businesses and industries are more or less dependent on particular localities for their reproduction (Cox & Mair, 1991). When local business interests are threatened by, for example, spatial restructuring in the wider economy, locally dependent business might respond to the threat by becoming more involved in local politics or by spreading risks across a wider geographical area. Such responses might include attempts to reorganize local government (for example, the creation of appointed rather than elected public agencies) and the securing of state powers and resources from higher (nonlocal) tiers of government. Viewed in this context, business involvement in downtown-orientated government coalitions is not the only strategic response to local accumulation and revenue crises; it is simply one of several policy alternatives.

Private business does not always rely on city government to bring about those institutional and territorial reorganizations deemed necessary for reinvigorating the local economy. When the business interest is more clearly defined at the metropolitan scale, the process of governing coalition formation will tend to take place at that scale and, consequently, policy choices and institutional reforms will be metropolitan-wide in scope and scale. Some business interests are more localized, however, and focus their coalition-building activities around local government, although in this case it might involve suburban cities. Residential developers, for example, depend on suburban housing markets and consequently tend to pursue more localized territorial reorganizations at the margins of metropolitan areas (Jonas, 1991). The actual process of governing coalition formation and the types of reorganizations pursued are therefore contingent on the territorial scale and scope of "local" business interests.

As urban regime theory recognizes, globalization has created severe problems of regulation for city government. But to reduce those problems to a division of labor between state and market is far too restrictive. Local governing capacity depends on the ability to regulate both the division of labor between state and market within the locality and to influence changes in wider territorial divisions of labor. A theory of strategic choice must therefore include the possibility that governing coalitions will attempt to seek reorganizations of scale divisions of labor in the state (Cox & Mair, 1991).

Writing in the British context, Painter and Goodwin (1995) have observed that attempts to regulate uneven development "usually involve, among other things, the development of a decentralized government tier" (p. 336). Once established that tier "is also subject to its own process of uneven development"

(p. 336). As an approach to understanding territorial reorganizations of the state, this is a useful start. But it needs to be more closely tied to a theory of structural context and strategic choice that takes explicitly into account spatial scale. As I have suggested, reorganizations within the state are contingent on territorial scope and scale of governing coalition activities and interests.

In the United States, there is considerable scope for territorial reorganizations at the local scale. In the broader context of globalization, metropolitan areas have increasingly been viewed as single, functional, economic regions that must exploit local comparative advantages to compete with other regions. Local actors with a regional stake in economic development have encouraged inward investment by seeking territorial reorganizations at a corresponding scale. "Competitive regionalism" has now become an important component of neoliberal urban policy (Cisneros, 1995), and is part of the context in which local governing coalitions make their policy choices.

Territorial reorganizations at the regional scale can, of course, be resisted by groups having a stake in existing local government structures. This is particularly true of suburban land development interests in the United States. Developers have typically sought to defend local (municipal and county) control of land use planning because it best serves their localized market interests. As a result, metropolitan areas tended to be marked by countervailing processes of political fragmentation and regional integration (Cox & Jonas, 1993; Hoch, 1984), although clearly globalization and the declining federal role in urban economies provide new scales of reference for these processes.

# Regulating Fordism: The Suburban-Defense Economy and Southern California

## The Suburban-Defense Economy and Urban Regime Transformations

The "classic" American model of fordism emerged from struggles in the Great Depression and took shape in the Cold War era following the closure of New Dealist policy experiments. It was based around the integration of mass production and mass consumption via Keynesian fiscal and regulatory measures. These measures established the "suburban-defense economy" as the dominant national MSR in U.S. fordism and sustained a period of productivity and wage growth that lasted approximately from the New Deal closure to the reorganization of the social-democratic welfare state, commencing in the Nixon and Carter administrations (Florida & Jonas, 1991). Thereafter American fordism entered into a sustained period of wage/productivity crisis that eventually encouraged a new wave of policy experimentation and institution building, as personified by President Reagan and his program of defense spending and monetarist fiscal policies. This program became a part of the wholesale "destructuring" of federal urban policy and continues with the Republicans' recent "contract with America," which targets many urban programs for budget cuts.

The postwar suburban-defense economy had four features that are salient for understanding the present period of crisis, restructuring, and political transformation under way in Southern California. First, this economy was built on a socially restrictive "class accord." In exchange for the right to unionize and engage in collective bargaining, core workers accepted employers' prerogatives to control production. Peripheral workers were excluded from the accord and did not enjoy the same standard of living as core workers, giving rise to segmented metropolitan labor and housing markets. Second, the fiscal advantages of suburban private home ownership were emphasized over renting and public housing assistance (Florida & Feldman, 1988), creating a dual housing market and reinforcing patterns of racial discrimination. Third, the federal government shifted its emphasis from a "full employment" policy to a more selective countercyclical job creation programs, including public works projects, freeway building, and urban renewal. Fourth, defense outlays and procurement were targeted away from traditional employment centers in the North East to new growth centers in the Sunbelt and West, including Southern California.

The suburban-defense economy provided a structural context for postwar urban regime transformations. The rise of New Dealist progrowth coalitions, which aligned the electoral interests of the Democratic party with the fiscal interests of central cities and downtown businesses, was associated with the impact of suburbanization on central-city tax bases (Mollenkopf, 1983). Later on, however, "white flight" and the suburban political backlash encouraged tax revolts and antigrowth sentiments in suburban areas. As suburban electoral interests began to dominate and the balance of power in Congress shifted, New Dealist progrowth coalitions fragmented. Progrowth interests continued to dominate Sunbelt city politics because cities were able to pursue aggressive annexation policies that enhanced the fiscal capacity of government and disenfranchised minority and suburban antigrowth interests (Fleischmann, 1986). More recently, however, Sunbelt city politics has experienced a convergence with Frostbelt urban politics (Kerstein, 1995) as a result of globalization, deindustrialization, and growing fiscal stress. Suburban areas in the Sunbelt are also experiencing economic problems as globalization processes penetrate throughout metropolitan areas.

## Territorial Dimensions of Suburban-Defense Destructuring in Southern California

Southern California has occupied a strategic position in the suburban-defense economy. As a focal point in the nation's "defense perimeter" (Markusen & Bloch, 1985), the region has received significantly high amounts of federal defense outlays and procurement per capita (Crump & Archer, 1993). Southern California has also sustained a tremendous rate of suburban land development (Fogelson, 1967; Kling, Olin, & Poster, 1991) thanks to the postwar federal policies and programs. The federal government has underwritten much of the region's infrastructure, and millions of Southern Californians have had access to federally insured mortgages. Through tax incentives and other programs, the government has subsidized the speculative ventures of the region's community builders and large-scale developers. An important feature of the suburban land development process in Southern California is its multijurisdictional character. Local services are provided by special districts and county governments, which contract with numerous local municipalities (Crouch & Dinerman, 1963; Miller, 1981).

Traditionally known for its citrus industry, the western San Bernardino-Riverside area—the Inland Empire—has historically occupied a peripheral location in the regional suburban-defense economy. For example, by far the majority of defense dollars have gone to contractors in Los Angeles and Orange counties, forming the basis of the growth of local industrial complexes around aerospace and related industries (Scott, 1988). During the 1980s, however, the Inland Empire became more integrated into the regional economy as suburban development spread eastward and industries relocated and expanded their operations. The region is the location of prominent defense contractors (for example, TRW's Ballistic Missiles Division) as well as two major military installations (Norton Air Force Base and March Air Force Base). The massive integrated steel mill operated by Kaiser Steel Corporation at Fontana was a major local employer until its closure in the early 1980s. The rise and demise of Kaiser's giant fordist enterprise, and the remaking of Fontana as a haven for homeowners fleeing crime, congestion, and underfunded schools in Los Angeles, provide keynotes for the Inland Empire's recent fortunes (Davis, 1990).

In the 1980s, San Bernardino-Riverside was the fastest growing metropolitan area in the United States. It grew in population from 1.6 million in 1980 to 2.6 million in 1990. Growth took the form of speculative middle- and low-income housing development, and was driven by housing price inflation in Los Angeles and Orange County. The Inland Empire provided cheap land and housing, and had access to major freeways corridors. Developers built homes in large quantities under minimal zoning restrictions, selling them for half the price of equivalent homes in Orange County and Los Angeles. Between 1983 and 1992, housing construction activity in the San Bernardino-Riverside SMSA kept pace with activity in Los Angeles County, although exceeding that of the Orange-Anaheim and Ventura-Oxnard SMSAs (see Figure 11.1). New home buyers traded proximity to their places of employment in Orange County and Los Angeles for an affordable suburban lifestyle. Citrus groves were rapidly transformed into tract housing developments, regional shopping malls, back office complexes, and a major edge-city commercial and manufacturing development around Ontario International Airport.

Toward the end of the 1980s, the contradictions of rapid growth and the overheating of the regional economy became apparent. Freeway congestion, sprawl, rising crime, and price inflation set in. City and county governments had underprovided services and infrastructure, and were fiscally stressed. Proposition 13 constrained local governments' revenue-raising powers, and municipalities and counties had to rely increasingly on unpopular sales taxes and user charges to support growth. During the 1980s, these growth problems were handled by existing local governments. More problematic was how to regulate the structural crisis that unfolded in the early 1990s.

The Inland Empire did not escape the effects of defense industry restructuring, although its direct local impact was initially muted. More serious was the indirect impact of job losses in Los Angeles and Orange County: Many workers employed in local defense establishments lived in the Inland Empire. The rate of unemployment in the Inland Empire rose dramatically, from 5.6% in 1987 to 11.0 % in 1992. This represented almost a twofold increase and was well above the regional average. There were also local closures, including the closure of Norton Air Force Base (A.F.B.) in San Bernardino County. Then in 1993, the federal Base Closure Commission announced its plans to deactivate and realign March A.F.B., with the prospect of further defense-related job losses.

The crisis had a dramatic impact on the Inland Empire housing market (see Figure 11.1). In 1989, residential and commercial construction activity fell sharply. The annual number of new private single-family housing unit permits issued in the two-county area (and mainly in western Riverside and San Bernardino counties) fell from 38,261 in 1989 to 12,366 in 1993, in other words more than a threefold decline. Falling demand put local developers and contractors out of business. Home prices also fell sharply, although not as sharply as in Los Angeles and Orange County. Thousands of homeowners were left in a

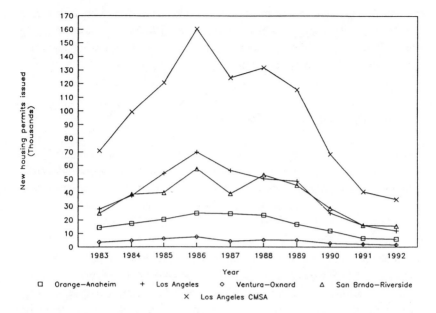

**Figure 11.1.** Housing Construction Activity in Greater Los Angeles Area, 1983 Through 1992
SOURCE: *California Statistical Abstract,* years shown.

financially precarious situation, and faced negative equity and foreclosure. Access to federally insured mortgages became more difficult, and that, together with declining consumption, exacerbated the structural crisis of the housing market.

# Regulating the Crisis: Strategic Capacities and Strategies in the Inland Empire

Local agencies have responded to the crisis in a variety of different ways, reflecting the complex and uneven nature of the suburban-defense transition as it has manifested itself in the Inland Empire. My focus here is on strategies for regulating suburban land development and local economic development as the region moves into a new growth phase and begins to see new rounds of inward investment. The crisis has called into question the institutions that governed the region's rapid growth phase during the 1980s. The argument I will make is that these institutions have not so much been replaced by new ones but rather have been strategically reorganized by local actors. In particular, there is an emerging emphasis on regional-level institutions as a means of regulating suburban politics during the current transition.

### Reregulating Suburban Land Development

The recent crisis has focused considerable attention on the sustainability of the suburban land development process. Given the importance of land development in the region's growth phase, attention has focused on the capacity of local government institutions to restore conditions conducive to future growth. Local control of land development has always been jealously guarded by local government in California, and frequent attempts to shift responsibility to the regional scale have faced intense opposition from local progrowth regimes (Pincetl, 1994). The recent crisis has, however, called into question the ability of local government to sustain growth, and has led to strategic interventions by progrowth interests at the municipal and county scales. Here I will focus on landowner and developer interventions in growth control measures, habitat conservation plans, and so-called development agreements.

The local agencies that promoted the rapid phase of suburban land development in the Inland Empire during the 1980s fitted many of the characteristics of the growth machine (Molotch, 1976), that is, they included speculator-developers, investors, and building and construction firms. County and city government for the most part supported this constellation of interests, providing the regulatory framework for land development, such as the preparation and modification of general and specific plans, enforcement of code specifications, and the completion of environmental impact reports (EIRs). Local planning departments were funded by developer fees, thereby making planning dependent on continuous land development, and limiting its use for progressive social goals. Unincorporated areas grew as fast as incorporated municipalities, and county government, which provided services to those areas, was an active player in the land development process.

When the housing market declined in the early 1990s, local planning agencies in the Inland Empire became chronically underfunded, forcing cities to lay off much of their planning staff. This tended to increase the influence of county government over land development activity because, unlike the cities, the counties could exploit economies of scale and retain adequate staffing levels. The activities of local progrowth regimes became more focused at the level of county where the resources lay and where important land use policy decisions were being made, although cities continued to have a stake in growth within their jurisdictions.

The growth machine theory hypothesizes that resistance to land development comes from residents concerned about the impact of unregulated growth on use values in the living place (Logan & Molotch, 1987). During the 1980s, throughout much of Southern California, rapid residential growth spawned a "homegrown revolution" (Davis, 1990) in which conservative homeowners' associations, concerned about the impacts of congestion, rising taxes, and white flight on property values, brought about restrictions to the encroachment and development of medium- and high-density housing tracts in inner suburban neighborhoods. Such restrictions helped push the frontier of suburban land development further eastward into western Riverside and San Bernardino counties. As a consequence, the Inland Empire began to attract medium- to high-density housing tracts, with local speculator-developers taking advantage of weak county development restrictions and codes. In Riverside, for example, county zoning categories allowed developers to concentrate single-family homes and apartments on relatively small lots having minimal services. As the pace of housing development activity increased during the 1980s, newly incorporated municipalities in the county were unable to provide adequate infrastructure, and were forced to impose developer impact fees to fund schools, libraries, and parks. Because housing prices were not as inflated in the Inland Empire as in other parts of Southern California, developers could pass the cost of impact fees onto new home buyers with little opposition or concern about its effect on demand.

Local antigrowth interests in the Inland Empire were politically unsophisticated and poorly funded. Growth control initiatives tended to be sporadic and issue or location specific. Nevertheless during the 1980s, growth control measures were introduced by several Inland Empire cities (and at the county level) in the midst of extraordinary growth (Warner & Molotch, 1995). In a few cases, growth in unincorporated areas led to movements for municipal incorporation as citizens tried to wrest control of land use planning from county government. These growth control measures placed restrictions on residential development (for example, preventing hill-slope development or limiting sewer hook-ups) but did not apply to commercial and industrial development.

Reviewing growth control measures in three subregions in Southern California (one of these being western Riverside County), Warner and Molotch (1995) grouped such measures into four categories: symbolic, episodic, countervailing, and counterinitiatives. Symbolic measures are those that politicians verbally support but undermine through actual policies. Episodic measures apply only in particular time contexts and can be negated by subsequent actions. Countervailing measures apply to some areas but encourage growth in others. Finally, counterinitiatives are taken by those subjected to growth control restrictions (i.e., landowners and developers). Landowner and developer counterinitiatives have been particularly effective in challenging growth control measures and ordinances in the Inland Empire. In the City of Moreno Valley (1990 population approximately 130,000), for example, speculator-developers financed a political campaign to defeat a citizens' growth control initiative (Measure J) in 1990. Measure J was voted down by a 3-to-1 margin.

In the Inland Empire, developer groups, such as local chapters of the Building and Industry Association (BIA) of Southern California, have bitterly fought against regional (multijurisdictional and statewide) growth control measures. This follows a historical pattern in California where Progressive-era government reforms have encouraged greater private interest group participation in the machinery of government, enabling special interest groups, such as business and industry organizations, to frustrate regional growth management initiatives (Pincetl, 1994). Existing regional institutions that might otherwise have assumed a growth control function have lacked the necessary powers and resources to control development on a regionwide basis. The functions of county Local Agency Formation Commissions (LAFCOs), for example, are limited to monitoring disputes over jurisdictional boundaries and service areas and establishing procedures for municipal annexations and incorporations. Regional Councils of Government (COGs) are multijurisdictional agencies that perform planning and advisory functions for member jurisdictions. Neither LAFCOs nor COGs have played a proactive role in growth control.

Perhaps more significantly, however, the crisis of the early 1990s brought an end to efforts by California Governor Pete Wilson's administration to develop a regional growth management plan at the state level. The crisis also seems to have diminished the incentive for growth control in the Inland Empire. In a telephone survey conducted in 1995, 20 out of 23 responding Inland Empire cities reported not having any current growth control restrictions. Only 2 cities reported experiencing high growth rates since 1990, 19 reported moderate to low growth, and 2 had experienced no growth at all since 1990.

Another area in which progrowth interests have been active at both the municipal and county levels is habitat conservation planning. Although conservation planning is not the same as growth control (even though landowners and developers often portray it as such), it does place restrictions on where growth can occur. To the extent that private property comes under those restrictions, conservation planning can be conflictual because it poses a potential threat to local growth interests and to the expansionary rights of landowners (Plotkin, 1987). Several regional conservation plans are being developed for sensitive habitats in Southern California. Arguably the most encompassing initiative is the statewide Natural Communities Conservation Plan (NCCP). Riverside and San Bernardino counties have developed their own conservation plans, however, and Riverside's plan has been especially controversial.

In 1988, a rare species of mammal unique to western Riverside County, the Stephens Kangaroo Rat (SKR), was listed as endangered and, under the provisions of the federal Endangered Species Act (ESA), triggered an immediate halt to all development in areas defined as being within the SKR habitat boundary. The listing prompted the County of Riverside to act as the lead agency and encourage local municipalities to enter into a voluntary planning agreement. Under the terms of Section 10 of the ESA, the county formed a habitat conservation agency to develop a habitat conservation plan (HCP) for the long-term protection of SKR habitat. By participating in the SKRHCP, cities and landowners with property known to contain SKR habitat were allocated "take" (i.e., destruction) of habitat in exchange for mitigation and entering into an agreement to acquire and set aside property for conservation reserves and study areas.

At an early stage in the planning process, county supervisors identified local "stakeholders" (individuals and agencies known to have an interest in the outcome of the plan), which were appointed to serve on an advisory committee. This committee was to make recommendations to the habitat conservation planning agency, in particular, regarding the size and location of reserves and study areas. The original purpose of the committee was to establish sound scientific guidelines for the design of reserve space, but it soon became a forum for progrowth interests to express their views, which for the most part were in line with those of the county supervisors who had made the appointments in the first place.

Leading the progrowth faction was the local chapter of the BIA, which believed that the goals of the conservation agency should not interfere with the interests of the housing industry and of member municipalities. This argument for local control was supported by the local Farm Bureau, which took a property rights stance. The BIA and the Farm Bureau formed a coalition of like-minded stakeholders, including landowners, developers, and an influential private property group. The coalition came up with its own recommendations that it brought to the advisory committee in the form of general guidelines for the conservation plan. Scientists and environmental groups serving on the advisory committee failed to make effective counterproposals, and, as a result, the recommendations of the progrowth coalition were incorporated into the final plan. These included eliminating scientific criteria for reserve design and severe changes to the size and locations of study areas and reserves. The final plan was limited mainly to lands under public ownership, thereby freeing private land from conservation restrictions.

San Bernardino County is in the process of drafting a multiple-species habitat conservation plan (MSHCP) for endangered or threatened species in that county. The draft plan involves an agreement between the county—the lead agency—and 15 cities. It states, among other things, that "individual landowners, groups of landowners, or development interests may choose to comply with the terms and conditions of the MSHCP affecting their proposed activities" (and apply for incidental take permits under the terms of the ESA) or, alternatively, developers and landowners can prepare their own conservation plans and apply for incidental take permits "outside the existing conservation plan umbrella" (San Bernardino County, n.d., p. 4). Like the Riverside plan, this conservation plan leaves considerable scope for suburban land development activity to continue in the county despite the threat of conservation restrictions.

County government has emerged from the conservation planning process as a strategically influential agency in the regulation of suburban land development in the Inland Empire. Although both Riverside and San Bernardino counties have experienced fiscal stress, the respective county governments have had both the capacity and resources to sustain land development during the recent crisis, and, consequently, progrowth interests have tended to focus their activities on the county rather than the municipality scale. This emerging role for county agencies reflects the fragmented form of suburban land development that has made it difficult for any single local government agency to regulate and control development.

Special districts and single-purpose authorities have proliferated in the unincorporated areas, many of which have protected the fiscal interests of county government while establishing conditions for a new round of development activity. A case in point is the creation of development agreements for specific planning areas within county unincorporated territories. Landowners and developers have entered into such agreements with county government to guarantee vested rights in future land development activities. In Orange County, development agreements have, in the absence of county funds, financed road improvements. In return for agreeing to subsidize road schemes, developers are given entitlements to develop in designated areas for as long as 20 years. Similar principles have been applied to development agreements in western Riverside County, many of which were negotiated after the listing of the SKR in 1988 and before the enforcement of a county growth-control measure. Areas contained in these development agreements that might otherwise have been subject to growth restrictions have begun to experience a great deal of land development activity.

## Local Economic Development and Regional Governance

Regime theorists have focused considerable attention on informal partnerships between downtown business elites and city government agencies (Cummings, 1988; Stone, 1989). Such partnerships have been responsible for the redevelopment and economic transformation of downtown areas throughout the United States (Judd & Parkinson, 1991; Leitner, 1990; Leitner & Garner, 1993). These transformations have occurred in the wider context of the restructuring of central cities from centers of manufacturing to postindustrial centers of corporate control and service functions. Industrial decentralization and service employment growth have also transformed suburban spaces, although these days the outermost suburbs tend to be recipients rather than losers of manufacturing firms. The growth of edge cities, industrial parks, and back-office complexes at the margins of metropolitan areas like Los Angeles bears witness to the process of industrial decentralization. Yet in comparison with regime theory's emphasis on central-city redevelopment regimes, regime formation around suburban redevelopment projects has received little attention (a possible exception is Kerstein, 1993).

Redevelopment activity in suburban areas is limited to a small scale because suburbs have not necessarily experienced the same level of economic decline as have central cities, and local governments are often lacking the fiscal and institutional capacities found in the larger central cities. Nevertheless redevelopment has become a very important institutional mechanism for local governments in suburban California, and, in the Inland Empire, has taken on new significance because of its economic development potential. That potential, however, has been frustrated by metropolitan fragmentation which tends to encourage interjurisdictional competition for inward investment.

Thus, with the aim of stimulating local economic development via inward investment, local actors (including county and city economic development and redevelopment officials, local mayors, and federal, state, and local politicians) have coalesced and introduced regional scale institutions of governance. Incentives for competitive regionalism have come from two directions: below and above. First, there have been bottom-up attempts to establish a regional economic partnership that consolidates and coordinates the activities of local economic development agencies throughout the Inland Empire. Second, there have been top-down federal incentives to organize regional partnerships as a means of competing for federal defense restructuring contracts and grants.

In California, redevelopment is a state program that allows local governments to raise local revenues for commercial redevelopment and affordable housing. It is recognized in the California constitution and funded by a local revenue source known as tax-increment financing (TIF). Redevelopment can be used by cities and counties throughout the state to combat urban blight.[1] The law also provides that redevelopment agencies set aside 20% of their tax increment to provide affordable housing. Although agencies are notorious for not spending their housing set-aside funds, redevelopment agencies can use redevelopment to secure matching CDBG funds from the U.S. Department of Housing and Urban Development.

California's community redevelopment law was enacted in the 1950s as a means of generating local matching funds for federal urban renewal projects.[2] Since its enactment, the law has been widely interpreted by city and county agencies. In the Inland Empire, for example, redevelopment is now used for a variety of purposes that are not strictly concerned with solving urban blight, including flood control, hotel and resort redevelopment, property acquisition, industrial and commercial development, business retention, downtown revitalization, historic district renovation and infrastructural improvements.

During the 1980s, the number of redevelopment projects under way in the Inland Empire dramatically increased. This increase was in response to the 1978 passage of California's property tax limitation measure, Proposition 13, which radically altered the state's tax structure. For cities and counties, it meant increasing reliance on nonproperty revenue sources, such as sales taxes and user fees, to fund general operations. Redevelopment became a means by which a city or a county could "create" a local tax base by subsidizing the development of revenue-generating land use activities, including shopping centers and auto malls. Almost every city in the Inland Empire wanted and now has a redevelopment agency (a total of 26 agencies). Annual tax increment to these agencies increased from just under $26 million in 1985 to more than $200 million in 1992-1993 (Figure 11.2).

One reason for the growing popularity of redevelopment is the possibility that local government and progrowth interests can develop partnerships for specific projects, both in downtown areas and other parts of the suburbs. Unlike city government, a redevelopment agency can engage in formal and informal transactions and negotiations with the private sector. Its activities are conducted separately from other parts of government, and agencies are not exposed to the same level of public scrutiny.

Interviews with redevelopment officials in the Inland Empire reveal that progrowth interests tend to dominate redevelopment regime activities, but progrowth *coalitions* form only when projects become controversial. Redevelopment agency officials do not enter into long-term partnerships with residential developers, and indeed many officials noted that developers have become less involved in progrowth activities since the 1980s. Officials increasingly rely on

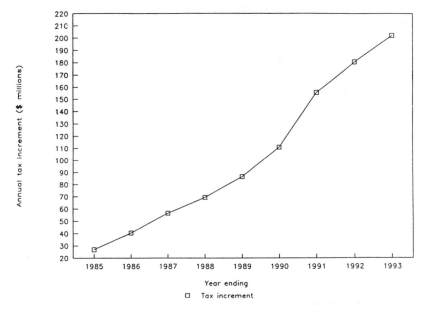

**Figure 11.2.** Annual Tax Increment Revenue to Inland Empire
Redevelopment Agencies, 1985 Through 1993
SOURCE: State of California, State Controller, *Annual Report of Financial Transactions Concerning
Community Redevelopment Agencies of California,* years shown.

an extensive network of contacts with chambers of commerce, banks, commercial and industrial real estate brokers, and city and county economic development agencies throughout the Inland Empire region to coordinate their redevelopment projects (see Althubaity, 1995).[3]

Many officials emphasize the economic development potential of redevelopment, and no longer see it as performing traditional slum clearance and urban renewal functions. Recent changes in California's redevelopment code have indeed emphasized its urban economic development function. The law now provides that (a) a majority (80%) of a redevelopment project area must already be "urbanized," and (b) agencies may use redevelopment for the purpose of economic development

The recent crisis has revealed the limited capacity for redevelopment to serve all of the region's economic development needs, however. Many agencies have failed to generate sufficient tax increment to cover initial investments.[4] These fiscal problems are compounded by interjurisdictional competition for high-profile projects, such as regional shopping malls and convention centers.

Redevelopment officials are in the process of reassessing the strategic importance of redevelopment, putting more emphasis on integrating redevelopment projects with regional economic development goals. This compares with the 1980s when redevelopment was mainly used as a growth-enhancing tool, contributing to the "fiscalization" of land use that followed the passage of Proposition 13.

The City of Ontario has used redevelopment to attract "just-in-time" warehouse and distribution operations to the Ontario International Airport area, and the City of Riverside has used it to attract a large food distribution facility to its industrial redevelopment project area on the city's east side. In addition to encouraging inward investment, redevelopment officials are involved in business retention. In this area, redevelopment activities have been coordinated with those of state and local (city) enterprise zones. When the Rohr Company, a producer of aircraft parts, recently announced its plans to relocate out-of-state its Riverside facility, employing some 600 workers (down from more than 3,000 in 1990), the City agreed to lower its utility charges and negotiated with the State to extend its enterprise zone to include Rohr's plant. Rohr has since agreed to expand its Riverside operations.

Before the crisis set in, Inland Empire redevelopment and economic development agencies had few incentives to coordinate their activities at a regional scale. The crisis has altered that situation for two reasons. First, local economic development agencies have fewer resources with which to compete for outside investments and have found intraregional competition to be costly and detrimental to the region's ability to recover. Second, uneven development has placed the Inland Empire in an opportune position to attract businesses away from Los Angeles and Orange County. These considerations have recently led to "bottom up" pressures from city and county economic development agencies to attempt to coordinate their activities at the regional scale.

In 1992, city and county agencies agreed to form a new regional agency, the Inland Empire Economic Partnership (IEEP). The IEEP developed in the wake of the Los Angeles rebellion as a two-county strategy to encourage Los Angeles area business executives to consider relocating operations to industrial and commercial sites in the Inland Empire (Woolley, 1992). The IEEP's working framework is that Los Angeles area companies need to have access to the area's markets and skilled labor and prefer to relocate within the area rather than out-of-state. In exchange for an annual fee, the IEEP provides member cities information about prospective leads. It is then up to the economic development officials of the individual cities to match this information with an appropriate package of incentives. The IEEP has thus not totally replaced the functions of local government and local economic development agencies but rather coordinates their economic development strategies on a regional basis. In effect, the "beggar-thy-city-neighbor" competition of the 1980s has become the "beggar-thy-region-neighbor" competition of the 1990s.[5]

In addition to bottom-up pressures, there have been top-down incentives to establish a regional partnership for the Inland Empire. The issue here has been the coordination of defense restructuring in the wake of job losses in aerospace and other defense industries, the closure of Norton A.F.B. in San Bernardino, and the announcement that March A.F.B. in Riverside County will be deactivated and realigned in 1996. The planned closure of Norton A.F.B. was initially seen as an opportunity by a local progrowth coalition in Riverside County to expand March A.F.B., but when that base was also added to the Pentagon's closure list it was transformed into a local anticlosure coalition, which fought for realignment rather than outright closure. The county and the cities of Riverside, Perris, and Moreno Valley have formed a Joint Powers Authority to manage the restructuring of March A.F.B. into part civilian and part military reserve uses.

These military base closure and realignment plans have recently been incorporated into a regional effort to coordinate defense restructuring and seek federal contracts and grants (San Bernardino/Riverside Counties, 1994). This regional effort grew out of a two-county strategy initiated in 1991 by Democratic Congressman George Brown (whose district lies in western San Bernardino County) and the Inland Empire Congressional Caucus. In 1992, the Caucus hosted the first Inland Empire Economic Summit, a loose association of representatives from industries, local governments, community organizations, and educational institutions. Through its five "Action Groups," the Summit is charting a plan for sustainable regional economic recovery, known as the Inland Empire Defense Conversion and Economic Recovery Planning Effort. The plan includes encouraging local agencies and business to seek grants from federal agencies, including the Technology Reinvestment Project, NASA, the Department of Energy and the Economic Development Administration (Christensen & Pallia, 1994).

# Conclusions

The growing emphasis on regional partnerships in the Inland Empire illustrates the strategic importance of scale in the current transition period. The kinds of regional partnerships emerging reflect the region's changing position in the national economy, and the ways in which local actors have chosen strategically to respond to that change. The economic future of the Inland Empire region is inextricably tied to the fortunes of Southern California as a whole in the global economy, but the nature of the local response will continue to reflect conditions unique to the region. These conditions include the failure of local government institutions to manage rapid growth in the 1980s. In the 1990s, county government and regional partnerships have taken on a more prominent role in regulating the region's suburban-defense transition. Because some of these reorganizations place the Inland Empire in a competitive relationship to Los Angeles and Orange County, it is too early to tell if they are capable of sustaining growth over the longer term or even mark a fundamental shift toward a new mode of social regulation.[6]

The theoretical and empirical emphasis of this chapter has been on the treatment of scale in regulation theory and urban regime theory. Having started out as theories directed at two separate scales, the national and the urban, respectively, there is evidence of a convergence of focus on local governance. The danger is, however, that in the process of converging, the two theoretical approaches will ignore the complex and often contradictory role of scale in recent economic and political transformations. If regulation theory is to become more than merely a way of providing a framework for contextualizing local governance, and if regime theory is to develop into a robust theory of urban politics, then more attention should be given to territorial reorganizations at different scales in the state.

These conceptual concerns are highlighted by the structural context of recent transformations in American urban politics and policy. The effects of economic globalization and the declining federal role are being felt not just in the downtowns of large central cities but also at the outermost margins of metropolitan areas. In this context, the question of governance is not so much one of "who governs cities?" as one of "at what spatial scale?"

# Notes

1. Redevelopment agencies are run by professional staffs with management or planning backgrounds. City councilors serve on the board of directors of agencies. Redevelopment agencies are usually located in a separate department of city government, and increasingly are subsumed under city economic development agencies. Redevelopment projects are funded by the tax increment with results from any increase in assessed valuation of property over and above the frozen valuation at the start of a project. Part of the tax increment is used to recover the debt on bonds issued by the redevelopment agency, part is used to acquire property, and most of the remainder is "passed through" to other taxing authorities, including school districts.

2. Inland Empire cities, including Riverside and San Bernardino, with established downtowns have used redevelopment for traditional "slum clearance and urban renewal" purposes, causing some

residential displacement. Riverside currently has a highly organized downtown progrowth redevelopment regime that has renovated the historic Mission Inn district and attracted some public functions away from the City of San Bernardino. Redevelopment has caused displacement of local homeless populations and brought the City into conflict with local neighborhood and parents' associations.

3. Based on personal interviews conducted with redevelopment agency officials in 1995.

4. Some redevelopment agencies that used TIF to underwrite growth in the 1980s are over-leveraged.

5. In 1993, IEEP activities led to 1,200 new jobs and $150 million in new investment in the Inland Empire (Inland Empire Economic Partnership, 1994). Local economic development officials estimate that only about 5% of investor leads come directly through the IEEP (from personal interviews conducted in 1995).

6. Regional organizations like the IEEP represent the trend toward "competitive regionalism" in U.S. urban policy.

# PART 4

# Regulating Urban Regimes

# 12

# Regulating Urban Regimes

*Reconstruction or Impasse?*

MICKEY LAURIA

In the introduction to this book, I offered regulation theory as providing a potential contextualizing approach that could assist in reconstructing urban regime theory, hence overcoming some of its inadequacies. The contributors to this collection have made attempts at providing a first cut at this integration. Although I did not suggest that this was to be a perfect marriage, I want to conclude this volume by evaluating the successes embodied in this book and the suggested avenues for further research and necessary theory development. But, first it will be helpful to rehearse the inadequacies and potential complementarities of urban regime and regulation theory.

Although urban regime theory focuses on the machination of political practices, it has been criticized for a tendency to equate the scale of the urban governance with municipal politics and politicians (Cox, 1991b). This criticism has always seemed misplaced to me, for in much regime analysis, researchers focus on local municipal politicians (the mayor) but also on state-level representatives, national congressmen, and other extralocal jurisdictional politicians (e.g., county and localized, geographically circumscribed, state boards) and their role in local governance. It should be clear that regime analysts do not, in my mind, equate local governance and urban regimes with the dominant municipality's politicians plus engaged economic actors. On the other hand, what this criticism makes clear is that regime analysts have excluded other suburban and exurban municipal politicians from the mix. Thus, regime analysts have tended to assume (not empirically evaluated) the dominant role of the central city's municipal politicians in the metropolitan governing coalition—often to the point of ignoring the role of other municipal politicians. This is a serious flaw, as Cox's focus on cooperation and transaction cost theory demonstrates. To compound this problem of inadequately articulating intralocal political machinations, urban regime theory has a further scale problem in that it has also inadequately theorized the connections between local agents (economic and political) and their wider institutional context.

Regulation theory is not without its shortcomings. Although regulation theory does focus on extralocal political and economic influences, it has traditionally underestimated the importance of local actors and organizations and thus has not been able to explain the concrete construction of regulatory mechanisms. Regulation theory also has inadequately conceptualized scale; its abstractions have often ignored historically significant spatial variations in material and discursive practices.

These scale problems suggest a more important ontological problem: urban regime theory tends to be tied to causal relations explained with rational choice theory whereas regulation theory is more open to causality being based on other than individual rational actions. But, because regulation theory is only beginning to grapple with material and discursive practices, it has yet to demonstrate how that broader theory of causality would operate. Thus, if urban regime theory is to provide the analysis of material and discursive practices, it must do so without resorting to purely rational behavioral processes. Both approaches undertheorize capitalism: Urban regime theory has no explicit theory short of the division between market and state whereas regulation theorists have often reduced capitalism's complexity to discrete transformations between homogeneous accumulation regimes that ignore how material practices constitute modes of regulation.

Their complementarities are important and promising. First, both have an overarching concern with governance: regime theory with political coalitions and their capacity to govern whereas regulation theory emphasizes the governance of production and consumption systems. An understanding of governance is necessarily incomplete without the exposition and articulation of these two components of a political economy. Second, their scale deficiencies complement rather than compound each other. It is possible that regulation theory's regulatory processes can be used to contextualize and articulate urban regime theory's local political economic machinations, and urban regime analysis can help embody the material and discursive social practices of regulatory processes.

Finally, although they both inadequately theorize dynamic capitalist social relations, their complementarities suggest that a theoretical reconstruction is possible by empirically focusing on the concrete social practices of urban politics in specific places and times.

These complementarities also suggest that this is not to be empiricist research, but rather reflexively previsioned with careful abstractions that attempt to resolve the aforementioned deficiencies of the urban regime and regulation theories. Thus, from within regulation theory's gaze, as urban regime analysts, we know to evaluate the local structure of capital and the various fractions of capital within the governing coalition, to look for emerging institutional arrangements tied to consensus seeking and social regulation, and to evaluate the external connections of local politicians and capitalists.

## Explaining the Regulation of Urban Politics: Habitus, Spatial Structures, or Hegemony

Painter, Feldman, and Jessop all offer somewhat different attempts to resolve the above mentioned ontological, epistemological, and subsequently methodological problem of fusing an urban regime theoretical approach with a regulationist approach. Painter suggests that a reconstructed urban regime theory may provide an approach that examines the mediation of regulation in and through specific social practices and forces, thereby resolving regulation theory's inability to nonfunctionally explain change. Compatible with my criticism of Stone's urban regime theory (Lauria & Whelan, 1995), Painter argues that urban regime theory, as now constituted, explains regime formation and transition based on rational choice theory of behavior and thus is incompatible with regulation theoretic accounts. Painter proposes replacing the rational choice theoretic base of urban regime theory with Bourdieu's conceptualization of *habitus, field,* and *strategy* as a way of understanding the processes by which potential participants in a regime come to join and interact within a governing coalition. He argues that urban politics can be seen both as a particular field with its own norms and habitus and as the intersection of different fields (government, business, voluntary, and civil society) where the different habitus and strategies collide, catalyze, and transform. Thus, clearly coalition formation must be seen as more than the constellation of discrete rational choices. Here, as seen in Feldman's chapter, methodologically an analysis of discursive practices becomes crucial in a reconstructed urban regime methodology. Finally, Painter's discussion of the habitus, field, and strategy of differing actors in a governing coalition and the subsequent stability or instability of the urban regimes closely resembles Jessop's turn toward Gramscian theory and the concept of hegemony (see also Clavel, 1995).

Jessop provides us with the basis for a reoriented urban regime research agenda. His interpretation draws on the complementarity of Gramscian state theory and regulation theory to contextualize urban regime analysis. From this basis, Jessop signals lessons to heed during the analysis of local economic governance. I will mention only three. He asserts that one needs to understand how the local economy comes to be constituted as an object of supralocal economic and extraeconomic regulation. Another lesson is to evaluate the relationship between local economic strategies and prevailing hegemonic projects. Third, he argues that it is important to analyze how institutions and apparatuses are strategically selective. Particular forms of economic and political systems privilege some strategies over others, some forces over others, and so forth. He argues that the durability of urban regimes depends on the coherence and economic feasibility of their strategies and on the strategic capacities rooted in local institutional structures and organizations, that is, how the urban regime is linked to the formation of a local hegemonic bloc and its associated historical bloc.

Feldman attempts to develop a theory of spatial structures necessary for understanding urban regime formation. He retheorizes regulation theory, taking into account the social effects of spatial organization. He develops a new conceptual tool, spatial structures of regulation (SSR), to situate urban regimes with respect to six circuits of capital that constitute a particular SSR. He argues that an analysis of a locality's SSR would help identify issues that are important to local capital and help explain the nature of their involvement in local politics. In the other direction, such analysis could help better understand the effects of local public policy. More important, Feldman argues that urban regime theory can help us understand the role of the local state in the social construction of the local SSR.

## Redressing Urban Regime Theory in a Global Economy

A theoretical problem identified with urban regime abstractions is that they have failed to account for the regulation of urban politics in a global economy. Although in different language, Leo evaluates the geographical variation in the role of the central state and regulatory processes and their interactions with the politics of urban planning and economic development. Leo criticizes regime theory for failing to locate urban politics in a national and global context. He argues that in cases where the central state takes a larger role in urban development processes (e.g., France, United Kingdom, and Canada), the power of large corporations is somewhat neutralized or at least "the playing field is more level." Stated differently, local governments left to their own (as often is the case in the United States) are more vulnerable to the ever increasing mobility of finance capital of large development projects and corporate capital of offices and branch plants. At the same time, Leo finds that this central state role also correlates with less citizen participation that he speculates to be of increasing importance in the future. Thus, for Leo urban regime theory needs to be reconstructed in a fashion that internalizes national and global political economics.

Whereas Leo's contribution to the reconstruction of urban regime concepts was to highlight differences in nation states' role in urban redevelopment, Cox focuses on the reconceptualization and specification of mechanisms of cooperation in the governance of urban development and the spatiality of that governance. In the former, he focuses on the insights provided by considering transaction cost theory insights to explain cooperation within and between state and market actors. In the latter focus, he highlights the erroneous effects of viewing space as a mere backdrop or arena for political economic relations and social relations in general to play within. Here he focuses on the concept of local dependence and its implications for cooperation and the transaction cost problem. Cox corrects the privileged role often given to local politicians and bureaucrats in coordinating the governance of local economic development strategies. He argues that such governance is more contingent than that recognized by urban regime analysts.

## Concrete Research: Regulating
## Urban Politics in a Global Economy

Horan argues that urban regime theory needs to and that regulation theory is useful for refocusing our gaze on the problem of politically bridging the gap between market and state in urban governance. She argues that although the two theoretical approaches have a common focus—the contingent nature of urban governance—the regulationist conception of governance involves the more complex task of social-political-economic harmonization than the urban regime idea of political coordination. At the same time, Mann's concept of infrastructural power focuses her attention on how local governments can contribute to spatial advantage. Thus, she evaluates the historical development of Boston's governing coalition in its market context, the shifting of the distribution of the tax burden, the reorganization of the local state administrative and legislative apparatuses, and their relationship to citizen participation and electoral coalitions.

Beauregard interprets the development of Philadelphia's postwar urban regime within the U.S. postwar regime of accumulation and mode of social regulation. He argues that Philadelphia's regime was not solely the result of local motivations, capacities, and cooperation but, rather, these were facilitated by a shift in the mode of regulation within capitalism. Subsequently, when that mode of regulation began to break down, Philadelphia's development-oriented regime began to unravel. Although the objective conditions that demanded reinvestment continued to worsen, publicly led reinvestment diminished. Beauregard argues that the postwar labor-capital accord that provided greater governmental involvement in social welfare activities was important in supporting local progrowth governing coalitions. Local government became an object and agent of this mode of social regulation. Its charge was to simultaneously facilitate and ameliorate the negative externalities of uneven spatial development through slum clearance, urban renewal and public housing, and welfare state services. As the fordist regime of accumulation began to falter and industrial restructuring weakened the labor-capital accord, the associated mode of social regulation became counterproductive. Thus, the federal government began a withdrawal from welfare state activities. This shrinkage of federally funded welfare state activities forced local governments to become more entrepreneurial to hold their position in the urban governing coalition. At the same time, global industrial restructuring began to weaken the ties of some capital to particular localities. Thus, the governing coalitions began to reorganize and regimes began to break down.

Keating challenges the claim that Cleveland is the "comeback" city, that is, that Cleveland reversed its decades of decline and is now revitalized. Keating focuses on intercity competition for professional sports franchises as but one aspect of cities being the last entrepreneurs. He argues that his analysis of Cleveland's governing coalition's strategy for downtown redevelopment epitomizes the insights provided by the urban regime approach: the primacy of public financing (local, state, and national) of large private development projects, the role of large economic interests in determining the growth agenda and form of specific projects, the ideological/consensus seeking role of local politicians in selling the projects to the public, the creation of quasi-public entities with minimal public accountability and oversight to plan and implement the development projects, and the lack of national reglementory intervention in local place-making competitions in the United States. Thus, the Cleveland case epitomizes the regulation of urban politics and urban development projects in a global economy where private capitals reduce economic risk, increase potential profits, and control the development arena and thus their competition through urban politics. Though Keating indicates that Cleveland is not unusual here, he also points to the contingent nature of this urban politics by referring to cases (San Francisco and Seattle) where the urban regime's electoral coalition was undermined via local initiative.

Jonas also focuses on the role of territorial scale (spatiality) in the regulation of urban governance. As were Feldman, Jessop, Goodwin and Painter, and Painter, Jonas is critical of urban regime and regulation theory's ambivalence toward scale and its causal properties. His analysis of the reorganization of local governance, particularly in relation to land development and redevelopment, demonstrates the constitutive causal power of geographical scale. Spatial scale played a role not only in political coalition building but also in state institution reformation and in the political coordination of economic development strategies. The urban politics in Southern California that Jonas describes may be one emerging political economic form in the postfordist regime of accumulation.

# Conclusion

How does a dialectic synthesis of urban regime theory and regulation theory help us better understand both urban politics and regulatory processes? First, whereas Leo argues that urban politics must be analyzed within the context of its nation-state and the global economy, Goodwin and Painter argue that regulation ought to be seen as not only national but also the purview of the local state. Not surprisingly, Leo's comparative analysis highlights Goodwin and Painter's, Jessop's, and Feldman's emphasis on the temporal and spatial variation of regulatory processes and uneven development outcomes. Thus, it should be clear that the political economy of cities and the regulation of the capitalist economy are best understood in relation to each other.

Second, a synthesis allows us to move beyond a rational choice theoretical explanation of political economy while providing us with a methodology that allows us to analyze the material and discursive practices of urban politics. The complementarity of a focus on political practices (urban regime theory) and extralocal and extraeconomic regulatory processes allows researchers to focus on how urban regimes (or at best growth coalitions) can orient local accumulation strategies to help position their local economic space in the urban hierarchy and the international division of labor. This strategic diversity in accumulation strategies should not be surprising. It helps explain "why both politics and economics matter" and why analysts have found spatial and temporal variation in regulatory processes that are constituted through material and discursive social practices.

Third, this synthesis unveils or gives broader meaning to the changing structure and arena of collective decisions and its regulatory and political significance. This "hollowing out of the nation-state," as Jessop calls it, and the corresponding local creation of quasi-public consensus-seeking para apparats concerned with economic development and insulated from democratic processes has altered the development of governing coalitions (altered their fields and constituting habitus and strategies) and constitutes dynamic social regulatory processes. The creation of these para apparats was clearly seen in each of the case studies. On the one hand, their creation was clearly strategic and meant to increase the local state's capacity to alter localities' position in the urban hierarchy and international division of labor. On the other hand, they have weakened the ability of local politicians in that decision making and potentially in the governing coalition while lessening the legitimacy of local government within the social regulatory process. This weakened position appears to signal the emergence of a new contradiction for urban social movements to lever with the corresponding effects on regime and regulatory transition.

Fourth, the synthesis highlights the importance of discourse, particularly the key role of economic discourses in framing accumulation strategies and hegemonic projects (see Wilson, 1995). Clearly the discursive framing of strategies and projects correspondingly creates differentiated material advantages and disadvantages within the governing coalition, helps structure electoral coalition strategies, and helps reconstitute regulatory processes.

Fifth, this dialectic synthesis provides analytic insight into both urban regime transition and the dynamics of evolving modes of social regulation (regulatory processes) and emerging regimes of accumulation. On the one hand, in Feldman's terms, an analysis of a locality's Spatial Structures of Regulation would help identify what issues are important to which local capitals and help explain their involvement or noninvolvement in the governing coalition. At the same time, it can help researchers understand the effects of public economic development strategies. Alternately, viewing regulatory processes and the discursive and material urban political practices embodied in urban regimes as mutually constituting provides a mechanism to explain urban regime transition (or at least transforming governing coalitions) and dynamic or emerging regulatory processes. This is clearly demonstrated in Beauregard's and Horan's representation of Philadelphia's and Boston's postwar regimes.

Finally, it should be clear that much theoretical work is still to be done. At the same time, our synthesis requires that this theoretical work must be accomplished through empirical analysis of discursive and material practices embodied within urban political economies. We are not yet at the point where the comparative analysis of existing case studies will provide much more theoretical insight. Although Leo's chapter demonstrated that this type of analysis was initially useful, there are few extant case studies theoretically driven by the fusion we propose.

# References

Abbott, C. (1983). *Portland: Planning, politics and growth in a twentieth century city*. Lincoln: University of Nebraska.

Abbott, C., Howe, D., & Adler, S. (1994). *Planning the Oregon way: A twenty-year evaluation*. Corvallis: Oregon State University.

Adams, C., Bartelt, D., Elesh, D., Goldstein, I., Kleniewski, N., & Yancey, W. (1991). *Philadelphia*. Philadelphia: Temple University Press.

Adde, L. (1969). Philadelphia. In *Nine Cities: The anatomy of downtown renewal*. Washington, DC: Urban Land Institute.

Aglietta, M. (1979). *A theory of capitalist regulation: The U.S. experience*. London: New Left Books.

Althubaity, A. (1995). *The role of tax increment financing in land use change and economic development in the Coachella Valley, Southern California*. Unpublished Ph.D. dissertation, University of California at Riverside.

Altschuler, A. A., & Gomez-Ibanez, J. A. (with Howitt, A. M.). (1993). *Regulation for revenue: The political economy of land use exactions*. Washington, DC: Brookings.

Anderson, P. (1991). *Imagined communities*. London: Verso.

Arrow, K. J. (1971). Economic welfare and the allocation of resources for invention. In *Essays in the theory of risk-bearing*. Princeton: Universities—National Bureau of Economic Research Conference Series.

Arthur D. Little Inc. (1958, May 9). *Research staff report to the downtown subcommittee of the Greater Boston Economic Study Committee*.

AuCoin, D. (1991, September 9). Mayor of neighborhoods faces dismay in the street. *Boston Globe*, p. 9.

Axelrod, R. (1984). *The evolution of cooperation*. New York: Basic Books.

Axford, N., & Pinch, S. (1994). Growth coalitions and local economic development strategy in southern England. *Political Geography 13*(4), 344-360.

Baade, R. A. (1996). Professional sports as catalysts for metropolitan economic development. *Journal of Urban Affairs, 18*(1), 1-18.

Baade, R. A., & Dye, R. F. (1988). Sports stadiums and area development: A critical review. *Economic Development Quarterly, 2*(3), 265-275.

Bacher, J. C. (1993). *Keeping to the marketplace: The evolution of Canadian housing policy.* Montreal: McGill-Queen's University Press.

Bacon, E. N. (1943). How city planning came to Philadelphia. *The American City, 58*(2), 62.

Baim, D. (1992). *The sports stadium as a municipal investment.* Westport, CT: Greenwood.

Bakshi, P., Goodwin, M., Painter, J. & Southern, A. (1995). Gender, race and class in the local welfare state. *Environment and Planning A, 27,* 1539-1554.

Bakvis, H. (1991). *Regional ministers: Power and influence in the Canadian cabinet.* Toronto: University of Toronto.

Baldassare, M. (1986). *Trouble in paradise: The suburban transformation in America.* New York: Columbia University Press.

Banfield, E. C. (1961). *Political influence.* New York: Free Press.

Barbrook, R. (1990). Mistranslations: Lipietz in London and Paris. *Science as Culture, 8,* 80-117.

Barlow, M. (1995). Greater Manchester: Conurbation complexity and local government structure. *Political Geography, 14*(4), 379-390.

Barnett, J., & Miller, N. (1983). Edmund Bacon: A Retrospective. *Planning, 49*(11), 4-11.

Bartimole, R. (1994). "If you build it," Baseball holds a city hostage. *The Progressive, 586,* 28-31.

Bartimole, R. (1995). Who governs: The corporate hand. In W. D. Keating, N. Krumholz, & D. C. Perry (Eds.), *Cleveland: A metropolitan reader* (pp. 161-174). Kent, OH: Kent State University Press.

Bauman, J. F. (1981). Visions of a postwar city. *Urbanism Past and Present, 6*(1), 1-11.

Bauman, J. F. (1987). *Public housing, race, and renewal.* Philadelphia: Temple University Press.

Beatty, J. (1992). *The rascal king: The life and times of James Michael Curley, 1874-1958.* Reading, MA: Addison-Wesley.

Beauregard, R. A. (1989a). Philadelphia: City profile. *Cities, 6*(4), 300-308.

Beauregard, R. A. (1989b). The spatial transformation of post-war Philadelphia. In R. A. Beauregard (Ed.), *Atop the urban hierarchy* (pp. 195-238). Totowa, NJ: Rowman & Littlefield.

Beauregard, R. A. (1993). *Voices of decline: The post-war fate of U.S. cities.* Oxford: Blackwell.

Benko, G., & Lipietz, A. (1994). De la régulation des espaces aux espaces de régulation. In R. Boyer & Y. Salliard (Eds.), *Théorie de la régulation: L'état des savoirs* (pp. 293-303). Paris: La Decouverte.

Bhaskar, R. (1979). *The possibility of naturalism.* Atlantic Highlands, NJ: Humanities Press.

Bier, T. (1995). Housing dynamics of the Cleveland area, 1950-2000. In W. D. Keating, N. Krumholz, & D. C. Perry (Eds.), *Cleveland: A metropolitan reader* (pp. 244-259). Kent, OH: Kent State University Press.

Blumenfeld, H. (1967). *The modern metropolis: Its origins, growth, characteristics, and planning.* Cambridge: MIT Press.

Body-Gendrot, S. N. (1987). Grass-roots mobilization in the Thirteenth Arrondissement. In C. N. Stone & H. T. Sanders (Eds.), *The politics of urban development.* Lawrence: University Press of Kansas.

Boston. (1978, May). *Boston comprehensive economic development strategy.*

Boston City Election Department. (Various years). *City Document No. 10.*

Boston Economic Development and Industrial Commission. (1992). *The missing link: A study of career opportunities for Boston residents in Boston hospitals and long-term care facilities.*

Boston Foundation. (1989, December). *In the midst of plenty.*

Boston Home Rule Commission. (1971). *Final report.*

Boston Redevelopment Authority. (1976, August 26). *Downtown Boston today.*

Boston Redevelopment Authority. (1992, February). *The economy.*

Bourdieu, P. (1990). *The logic of practice.* Cambridge: Polity. (First published in French under the title *Le sens practique* in 1980)

Bourdieu, P. (1994). Rethinking the state: Genesis and structure of the bureaucratic field. *Sociological Theory 12* (1), 1-18.

Bourdieu, P., & Wacquant, L. J. D. (1992). *An invitation to reflexive sociology.* Cambridge: Polity.

Boyer, R. (1986). *La théorie de la régulation: Une analyse critique.* Paris: La Decouverte.

Boyer, R. (1990). *The regulation school: A critical introduction.* (C. Charney, Trans.). New York: Columbia University Press.

Boyer, R., & Hollingsworth, R. J. (Eds.). (1995). *The embeddedness of capitalist institutions.* Oxford, UK: Oxford University Press.

Brenner, R., & Glick, M. (1991). The regulation approach: Theory and history. *New Left Review, 188,* 45-119.

Brown, J. (1986, August). *The revitalization of downtown Boston: History, assessment and case studies.* Boston: Boston Redevelopment Authority.

Burawoy, M. (1985). *The politics of production.* London: Verso.

Calaveta, N. (1992). Growth machines and ballot box planning: The San Diego case. *Journal of Urban Affairs, 14,* 1-24.

Calhoun, C., LiPuma, E., & Postone, M. (Eds.). (1993). *Bourdieu: Critical perspectives.* Chicago: University Press of Chicago.

Cappellin, R. (1992). Theories of local endogenous development and international cooperation. In M. Tykkulainen (Ed.), *Development issues and strategies in the new Europe.* Aldershot, UK: Avebury.

Carchedi, G. (1977). *On the economic identification of social classes.* London: Routledge Kegan Paul.

Carlaw, C. (1979, May). *A decade of development in Boston.* Boston: Boston Redevelopment Authority.

Carver, H. (1948). *Houses for Canadians.* Toronto: University of Toronto Press.

Castells, M. (1991). *The informational city.* Cambridge: Basil Blackwell.

Chamber of Commerce of Greater Philadelphia. (1951). *Philadelphia Facts 1951.* Philadelphia: Author.

Chandler, M. O. (1995). Politics and the development of public housing. In W. D. Keating, N. Krumholz, & D. C. Perry (Eds.), *Cleveland: A metropolitan reader* (pp. 228-243). Kent, OH: Kent State University Press.

Chen, X., & Orum, A. (1994, July). *City-building in the USA and China: An historical and comparative analysis.* Paper presented at the conference Shaping the Urban Future: International Perspectives and Exchanges. Bristol, University of Bristol, School for Advanced Urban Studies.

Chouinard, V. (1990). The uneven development of capitalist states: 1. Theoretical proposals and an analysis of post-war changes in Canada's assisted housing programmes. *Environment and Planning, A*(22), 1291-1308.

Christensen, B., & Pallia, D. (1994). *1993-1994 Report.* San Bernardino, CA: Inland Empire Economic Summit Defense Reinvestment Action Group.

Church, A., & Reid, P. (1995). Transfrontier cooperation, spatial development strategies, and the emergence of a new scale of regulation: The Anglo-French border. *Regional Studies, 29*(3), 297-306.

Cisneros, H. G. (1995). *Urban entrepreneurialism and national economic growth.* Washington, DC: U.S. Department of Housing and Urban Development.

Clark, G. (1981a). The employment relation and spatial division of labor: A hypothesis. *Annals of the Association of American Geographers, 71,* 412-24.

Clark, G. (1981b). Regional economic systems, spatial interdependence and the role of money. In J. Rees, G. Hewings, & H. Stafford (Eds.), *Industrial location and regional systems.* New York: J. F. Bergin.

Clark, J. S., Jr., & Clark, D. J. (1982). Rally and relapse: 1946-1968. In R. F. Weigley (Ed.), *Philadelphia: A 300-year history* (pp. 649-703). New York: Norton.

Clarke, S. E. (1987). More autonomous policy orientations: An analytical framework. In C. N. Stone & H. T. Sanders (Eds.), *The politics of urban development* (pp. 105-124). Lawrence: University Press of Kansas.

Clarke, S. E. (1988). Overaccumulation, class struggle and the regulation approach. *Capital and Class, 36,* 59-92.

Clarke, S. E. (1990). New utopias for old: Fordist dreams and post-fordist fantasies. *Capital and Class, 42,* 131-155.

Clavel, P. (1986). *The progressive city.* New Brunswick, NJ: Rutgers University Press.

Clavel, P. (1995). Regimes and progressive coalitions in cities. *Planning Theory, 14,* 44-64.

Cleveland City Planning Commission. (1988). *Cleveland civic vision 2000: Downtown plan.* Author.

Cleveland City Planning Commission. (1991). *Cleveland civic vision 2000: Citywide plan.* Author.

Cochrane, A. (1991). The changing state of local government: Restructuring for the 1990s. *Public Administration, 69,* 281-302.

Collier, R. W. (1974). *Contemporary cathedrals: Large-scale developments in Canadian cities.* Montreal: Harvest House.

Columbus annexation policy pays off. (1979, June 21). *Columbus Dispatch,* p. B3.

Coulton, C. J., & Chow, J. (1995). The impact of poverty on Cleveland neighborhoods. In W. D. Keating, N. Krumholz, & D. C. Perry (Eds.), *Cleveland: A metropolitan reader* (pp. 202-227). Kent, OH: Kent State University Press.

Cox, K. R. (1991a). The abstract, the concrete, and the argument in the new urban politics. *Journal of Urban Affairs, 13*(3), 299-306.

Cox, K. R. (1991b). Questions of abstraction in studies in the new urban politics. *Journal of Urban Affairs, 13*(3), 267-280.

Cox, K. R. (1993). The local and the global in the new urban politics: A critical view. *Environment and Planning D: Society and Space, 11,* 433-448.

Cox, K. R., & Jonas, A. E. G. (1993). Urban development, collective consumption and the politics of metropolitan fragmentation. *Political Geography, 12,* 8-37.

Cox, K. R., & Mair, A. J. (1988). Locality and community in the politics of local economic development. *Annals of the Association of American Geographers, 78,* 307-325.

Cox, K. R., & Mair, A. J. (1991). From localised social structures to localities as agents. *Environment and Planning A, 23,* 197-213.

Crouch, W. W., & Dinerman, B. (1963). *Southern California metropolis.* Berkeley and Los Angeles: University of California Press.

Crow, G. (1989). The use of the concept of strategy in recent sociological literature. *Sociology, 23*(1), 1-24.

Crump, J. R., & Archer, J. C. (1993). Spatial and temporal variability in the geography of American defense outlays. *Political Geography, 12,* 38-63.

Cullingworth, J. B. (1987). *Urban and regional planning in Canada.* New Brunswick, NJ: Transaction.

Cummings, S. (Ed.). (1988). *Business elites and urban development: Case studies and critical perspectives.* Albany: State University of New York Press.

Cutler, W. W. III. (1980). The persistent dualism: Centralization and decentralization in Philadelphia, 1854-1975. In W. W. Cutler III & H. Gillette, Jr. (Eds.), *The divided metropolis* (pp. 249-284). Westport, CT: Greenwood.

Cybriwsky, R. A., & Western, J. (1982). Revitalizing downtowns: By whom and for whom? In D. T. Herbert & R. J. Johnston (Eds.), *Geography and the urban environment* (pp. 343-365). London: John Wiley.

Dahl, R. (1961). *Who governs?* New Haven, CT: Yale University Press.

Davis, M. (1990). *City of quartz: Excavating the future in Los Angeles.* London: Verso.

Davis, O., & Whinston, A. (1965). Economic problems in urban renewal. In E. S. Phelps (Ed.), *Private wants and public needs* (pp. 140-153). New York: Norton.

DeLeon, R. E. (1992). *Left coast city: Progressive politics in San Francisco, 1975-1991.* Lawrence: University Press of Kansas.

Demirovic, A. (1988). *Regulation: Kollective praxis und intellectuelle.* Unpublished manuscript.

Devine, K. (1989, April 7). Neighborhood empowerment. *Charlestown Ledger.*

DiGaetano, A. (1989). Urban political regime formation: A study in contrast. *Journal of Urban Affairs, 11*(3), 261-283.

DiGaetano, A., & Klemanski, J. S. (1993a). Urban regime capacity: A comparison of Birmingham, England, and Detroit, Michigan. *Journal of Urban Affairs, 15*(4), 367-384.

DiGaetano, A., & Klemanski, J. S. (1993b). Urban regimes in comparative perspective: The politics of urban development in Britain. *Urban Affairs Quarterly, 29*(1), 54-83.

DiGaetano, A., & Klemanski, J. S. (1994, July). *Comparative urban regimes: Regime formation and re-formation in the UK and the US.* Paper presented at the conference Shaping the Urban Future: International Perspectives and Exchanges, Bristol, University of Bristol, School for Advanced Urban Studies.

Downs, A. (1988). The real problem with suburban anti-growth policies. *Brookings Review, 6*(2), 23-29.

Dreier, P., & Ehrlich, B. (1991). Downtown development and urban reform: The politics of Boston's linkage policy. *Urban Affairs Quarterly, 26*(3), 354-373.

Duchacek, I. D. (1984). The international dimension of sub-national self-government. *Publius, 14*(1), 5-31.

Dulong, R. (1978). *Les régions, l'état et la société locale.* Paris: Presses Universitaires de France.

Duncan, S., & Savage, M. (1991). New perspectives on the locality debate. *Environment and Planning A, 23,* 155-164.

Edwards, R. (1979). *Contested terrain: The transformation of the workplace in the twentieth century.* New York: Basic Books.

Ehrlich, B. (1987). *The politics of economic development planning: Boston in the 1980s.* Unpublished M.C.P. thesis, Massachusetts Institute of Technology.

Elkin, S. L. (1985). Twentieth century urban regimes. *Journal of Urban Affairs, 5,* 11-27.

Elkin, S. L. (1987). *City and regime in the American republic.* Chicago: University of Chicago Press.

Epstein, E. (1995, December 22). The Giants' latest pitch. *San Francisco Chronicle,* p. A-1.

Erie, S. P. (1988). *Rainbow's end, Irish-Americans and the dilemmas of urban machine politics.* Berkeley: University of California Press.

Esser J., & Hirsch, J. (1989). The crisis of fordism and the dimensions of a post-fordist regional and urban structure. *International Journal of Urban and Regional Research, 13*(3), 417-437.

Estall, R. C. (1966). *New England, a study in regional adjustment.* New York: Praeger.

Euchner, C. (1993). *Playing the field: Why sports teams move and cities fight to keep them.* Baltimore: Johns Hopkins University Press.

Fainstein, N. I., & Fainstein, S. S. (1983). Regime strategies, communal resistance, and economic forces. In S. S. Fainstein & N. I. Fainstein (Eds.), *Restructuring the city* (pp. 245-282). New York: Longman.

Fainstein, S. S. (1990). Economics, politics and development policy: The convergence of New York and London. *International Journal of Urban and Regional Research, 14*(4), 553-575.

Fainstein, S. S. (1994). *The city builders: Property, politics, and planning in London and New York.* Cambridge, MA: Blackwell.

Fainstein, S. S. (1995). Politics, economics and planning: Why urban regimes matter. *Planning Theory, 14,* 34-43.

References                                                                    247

Feagin, J. R. (1985). The social costs of Houston's growth: A Sunbelt boomtown re-examined. *International Journal of Urban and Regional Research, 9*(2), 164-185.

Feldman, M. M. A. (1977). A contribution to the critique of urban political economy: The journey to work. *Antipode, 9,* 30-49.

Feldman, M. M. A. (1995). Regime and regulation in substantive planning theory. *Planning Theory, 14,* 65-78.

Feldman, M. M. A., & Florida, R. L. (1990). Economic restructuring and the changing role of the state in U.S. housing. In W. van Vliet– – & J. van Weesep (Eds.), *Government and housing: Developments in seven countries* (Vol. 36, *Urban Affairs Annual Reviews*). Newbury Park, CA: Sage.

Feldman, M. M. A., & McIntyre, R. (1994). *Flexible production: Its incidence and implications for labor markets and economic development strategy* (Final Report to the Economic Development Administration, U.S. Department of Commerce). Washington, DC: National Technical Information Service.

Feldman, T. (1995). *Local solutions to land use conflict under the endangered species act: Habitat conservation planning in Riverside County.* Unpublished Ph.D. dissertation. University of California at Riverside.

Ferrer, A., & Van Til, J. (1994, July). *Barcelona and Philadelphia: An unlikely comparison.* Paper presented at the conference Shaping the Urban Future: International Perspectives and Exchanges, Bristol, University of Bristol, School for Advanced Urban Studies.

Finance Commission of the City of Boston. (1979, July 18). *The administration of Massachusetts general law chapter 121A by the City of Boston and the Boston Redevelopment Authority.*

Fleischmann, A. (1986). The goals and strategies of local boundary changes: Government organization or private gain? *Journal of Urban Affairs, 8,* 63-75.

Florida, R. L., & Feldman, M. (1988). Housing in U.S. fordism. *International Journal of Urban and Regional Research, 12,* 187-210.

Florida, R. L., & Jonas, A. E. G. (1991). U.S. urban policy: The post-war state and capitalist regulation. *Antipode, 23*(4), 349-384.

Florida, R. L., & Kenney, M. (1988). Venture capital, high technology and regional development. *Regional Studies, 22,* 33-48.

Fogelson, R. M. (1967). *The fragmented metropolis.* Cambridge, MA: Harvard University Press.

Foley, A. (1990, November 8). Metropolitan Boston employment structure. In *Technical appendix to the Howell Report* (Vol. 2). Boston: Boston Redevelopment Authority.

Forbes. (1995, October 16). The four hundred richest people in America. *Forbes, 1569,* pp. 262, 295.

Frieden, B., & Sagalyn, L. (1989). *Downtown Inc., how America rebuilds cities.* Cambridge: Massachusetts Institute of Technology.

Frye, N. (1982). *Divisions on a ground: Essays on Canadian culture.* Toronto: Anansi.

Garber, J. A., & Imbroscio, D. (1992). *Growth policies in Canadian and American cities: The myth of the North American city reconsidered.* Charlottetown, PEI: Canadian Political Science Association.

Gelfand, M. I. (1975). *A nation of cities.* New York: Oxford University Press.

Giddens, A. (1984). *The constitution of society.* Berkeley and Los Angeles: University of California Press.

Gillispie, M. (1996, February 3). Tribe, Cavs to cover costs of running sports complex. *Cleveland Plain Dealer,* p. B-6.

Goetz, E. G. (1991). Promoting low-income housing through innovations in land use regulations. *Journal of Urban Affairs, 13*(3), 337-351.

Goldberg, M. A., & Mercer, J. (1986). *The myth of the North American city: Continentalism challenged.* Vancouver: University of British Columbia.

Goldsmith, W. W., & Blakely, E. J. (1992). *Separate societies: Poverty and inequality in U.S. cities.* Philadelphia: Temple University Press.

Goodwin, M. (1992). The changing local state. In P. Cloke (Ed.), *Policy and change in Thatcher's Britain: A critical perspective* (pp. 77-96). Oxford, UK: Pergamon.

Goodwin, M., Duncan, S., & Halford, S. (1993). Regulation theory, the local state and the transition of urban politics. *Environment and Planning D: Society and Space, 11*(1), 67-88.

Goodwin, M., & Painter, J. (in press). Local governance, the crises of fordism and the changing geographies of regulation. *Transactions of the Institute of British Geographers, 21.*

Gottdiener, M. (1983). Some theoretical issues in growth control analysis. *Urban Affairs Quarterly, 18*(4), 565-569.

Grabowski, J. J. (1992). *Sports in Cleveland: An illustrated history.* Bloomington: Indiana University Press.

Gramsci, A. (1971). *Selections from the prison notebooks.* London: Lawrence & Wishart.

Gramsci, A. (1995). *Further selections from the prison notebooks.* London: Lawrence & Wishart.

Granatstein, J. L. (1971). *Marlborough marathon.* Toronto: A M Hakkert.

Granovetter, M. (1985). Economic action and social structure: The problem of embeddedness. *American Journal of Sociology, 91,* 481-510.

Greater Boston Economic Study Committee. (1959). *A report on downtown Boston.*

Hall, F., & McIntyre Hall, L. (1993/1994). A growth machine for those who count. *Critical Sociology, 20*(1), 79-102.

Hambleton, R. (1994, July). *Reinventing local government: A cross-national analysis.* Paper presented at the conference Shaping the Urban Future: International Perspectives and Exchanges, Bristol, University of Bristol, School for Advanced Urban Studies.

Harding, A. (1994). Urban regimes and growth machines: Towards a cross-national research agenda. *Urban Affairs Quarterly, 29*(3) 356-382.

Harding, A. (1995). Elite theory and growth machines. In D. Judge, G. Stoker, & H. Wolman (Eds.), *Theories of urban politics* (pp. 35-53). London: Sage.

Harvey, D. (1982). *The limits to capital.* Chicago: University of Chicago Press.

Harvey, D. (1985). *The urbanization of capital: Studies in the history and theory of capitalist urbanization.* Oxford, UK: Basil Blackwell.

Harvey, D. (1989). *The condition of postmodernity: An enquiry into the origins of cultural change.* Oxford, UK: Basil Blackwell.

Häusler, J., & Hirsch, J. (1987). Regulation und Parteien im Uebergang zum post-fordismus. *Das Argument, 165,* 651-671.

Hay, C., & Jessop, B. (1995, August 31). *The governance of local economic development and the development of local economic governance: A strategic-relational approach.* Paper presented at the annual meeting of the American Political Science Association, Chicago.

Heider, T. (1995, December 28). Gateway gets 10-year loan from county. *Cleveland Plain Dealer,* p. A-1.

Hepworth, M. (1990). *Geography of the information economy.* New York: Guilford.

Herrero, T. R. (1991). Housing linkage: Will it play a role in the 1990s? *Journal of Urban Affairs, 13*(1), 1-19.

Hill, E. W. (1995). The Cleveland economy: A case study of economic restructuring. In W. D. Keating, N. Krumholz, & D. C. Perry (Eds.), *Cleveland: A metropolitan reader* (pp. 53-88). Kent, OH: Kent State University Press.

Hirzel, D. M. (1993). Cleveland gateway. *Urban Land, 52,* 34-37.

Hoch, C. (1984). City limits: Municipal boundary formation and class segregation. In W. K. Tabb & L. Sawers (Eds.), *Marxism and the metropolis* (pp. 298-322). New York: Oxford University Press.

Hodgkinson, H. D. (1972). Miracle in Boston. *Massachusetts Historical Society, Proceedings, 84,* 71-81.

Home builders fear annexation. (1986, March 20). *Columbus Dispatch,* p. 2C.

Horan, C. (1990). Organizing the "new Boston": Growth policy, governing coalitions and tax reform. *Polity, 22*(3), 489-510.

Horan, C. (1991). Beyond governing coalitions: Analyzing urban regimes in the 1990s. *Journal of Urban Affairs, 13*(2), 119-135.

Horowitz, G. (1978, June 11). Notes on "conservatism, liberalism and socialism in Canada." *Canadian Journal of Political Science.*

Hoxworth, D. H., & Thomas, J. C. (1993). Economic development decision making in a fragmented polity: Convention center expansion in Kansas City. *Journal of Urban Affairs, 15,* 275-292.

Hunter, F. (1953). *Community power structure.* Chapel Hill: University of North Carolina Press.

Inland Empire Economic Partnership. (1994). *1993 achievements in economic development.* Ontario, CA: Inland Empire Economic Partnership, mimeo.

Isin, E. F. (1992). *Cities without citizens: Modernity of the city as a corporation.* Montreal: Black Rose.

Jackson, P. (1991). Mapping meanings: A cultural critique of locality studies. *Environment and Planning A, 23,* 215-228.

Jauhiainen, J. (1994, July). *A comparative framework in the question of waterfront redevelopment in Barcelona, Cardiff and Genoa.* Paper presented at the conference Shaping the Urban Future: International Perspectives and Exchanges, Bristol, University of Bristol, School for Advanced Urban Studies.

Jenkins, R. (1992). *Pierre Bourdieu.* London: Routledge.

Jennings, J. (1986). Urban machinism and the black voter: The Kevin White years. In J. Jennings & M. King (Eds.), *From access to power, black politics in Boston* (pp. 57-88). Rochester, VT: Schenkman.

Jenson, J. (1990). Representations in crisis: The roots of Canada's permeable fordism. *Canadian Journal of Political Science, 24*(3), 653-683.

Jenson, J. (1993). Naming nations: Making nationalist claims in Canadian public discourse. *Canadian review of sociology and anthropology, 30*(3), 337-350.

Jessop, B. (1982). *The capitalist state: Marxist theories and methods.* Oxford: Martin Robertson.

Jessop, B. (1983). Accumulation strategies, state forms, and hegemonic projects. *Kapitalistate, 10,* 89-111.

Jessop, B. (1990a). Regulation theories in retrospect and prospect. *Economy and Society, 19*(2), 153-216.

Jessop, B. (1990b). *State theory: Putting the capitalist state in its place.* Cambridge: Polity.

Jessop, B. (1992a). Fordism and post-fordism: A critical reformulation. In M. Storper & A. Scott (Eds.), *Pathways to industrialization and regional development* (pp. 43-65). London: Routledge & Kegan Paul.

Jessop, B. (1992b). Regulation und politik: Integrale Ökonomie und integraler Staat. In A. Demirovic, H.-P. Krebs, & T. Sablowski (Eds.), *Hegemonie und staat* (pp. 232-262). Münster: Westfälisches Dampfboot Verlag.

Jessop, B. (1993). Towards a Schumpeterian workfare state? Preliminary remarks on post-fordist political economy. *Studies in Political Economy, 40,* 7-39.

Jessop, B. (1994). Post-fordism and the state. In A. Amin (Ed.), *Post-fordism: A reader* (pp. 251-279). Oxford, UK: Blackwell.

Jessop, B. (1995a). The nation-state: Erosion or reorganization? In *Lancaster Regionalism Group working papers* (Governance Series, No. 50). Lancaster, UK: Lancaster University.

Jessop, B. (1995b). The regulation approach, governance and post-fordism: Alternative perspectives on economic and political change? *Economy and Society, 24*(3), 307-333.

Jessop, B. (1995c). Towards a Schumpeterian workfare regime in Britain? Reflections on regulation, governance and welfare state. *Environment and planning A, 27,*(6), 1613-1626.

Jessop, B. (1996a). The governance of uneven development: The East Thames corridor. *Lancaster Regionalism Group Working Papers (Governance Series), 54.* Lancaster, UK: Lancaster University.

Jessop, B. (1996b). Interpretive sociology and the dialectic of structure and agency: Reflections on Holmwood and Stewart's explanation and social theory. *Theory, Culture and Society, 13*(1), 119-126.

Jezierski, L. (1994, July). *Comparative partnerships and local regimes.* Paper presented at the conference Shaping the Urban Future: International Perspectives and Exchanges, Bristol, University of Bristol, School for Advanced Urban Studies.

Jonas, A. E. G. (1991). Urban growth coalitions and urban development policy: Postwar growth and the politics of annexation in metropolitan Columbus. *Urban Geography, 12,* 197-226.

Jonas, A. E. G. (in press). In search of order: Traditional business reformism and the crisis of neoliberalism in Massachusetts. *Transactions of the Institute of British Geographers, 21.*

Jonas, A. E. G. (1996). Local labour control regimes: Uneven development and the social regulation of production. *Regional Studies 30*(4), 323-338.

Jones, B., & Bachelor, L. W. (1986). *The sustaining hand: Community leadership and corporate power.* Lawrence: University of Kansas Press.

Judd, D., & Parkinson, M. (Eds.). (1991). *Leadership and urban regeneration: Cities in North America and Europe.* London: Sage.

Kantor, P. (1987). The dependent city: The changing political economy of urban economic development in the United States. *Urban Affairs Quarterly, 22*(4), 493-520.

Kantor, P. (1988). *The dependent city.* Glenview, IL: Scott, Foresman.

Katz, J. B. (1978). *The impact of federal grants in Boston in 1978.* Waltham, MA: Brandeis University.

Keating, W. D., & Krumholz, N. (1991). Downtown plans of the 1980s: A case for more equity in the 1990s. *Journal of the American Planning Association, 57,* 136-152.

Keating, W. D., Krumholz, N., & Metzger, J. (1989). Cleveland: Post-populist public-private partnerships. In G. D. Squires (Ed.), *Unequal partnerships: The political economy of urban redevelopment in postwar America* (pp. 121-141). New Brunswick, NJ: Rutgers University Press.

Keating, W. D., Krumholz, N., & Perry, D. C. (1995). The ninety-year war over public power in Cleveland. In W. D. Keating, N. Krumholz, & D. C. Perry (Eds.), *Cleveland: A metropolitan reader* (pp. 137-154). Kent, OH: Kent State University Press.

Keil, R. (1993). *Weltstadt—Stadt der Welt.* Muenster: Westfaelisches Dampfboot.

Kennedy, L. (1992). *Planning the city upon a hill, Boston since 1636.* Amherst: University of Massachusetts.

Kerstein, R. (1993). Suburban growth politics in Hillsborough County: Growth management and political regimes. *Social Science Quarterly, 74,* 614-630.

Kerstein, R. (1995). Political exceptionalism in sunbelt cities. *Journal of Urban Affairs, 17,* 143-163.

Keyes, L. (1969). *The rehabilitation planning game, a study in the diversity of neighborhoods.* Boston: Massachusetts Institute of Technology.

King, J. (1990). How the BRA got some respect. *Planning, 56*(5), 4-9.

King, M. (1981). *Chain of change, struggles for black community development.* Boston: South End.

Kissling, C. (1992, January 19). Gateway's "50-50" leans heavier on the public. *Cleveland Plain Dealer,* p. B-1.

Klaassen, L. H., & Cheshire, P. C. (1993). Urban analysis across the Atlantic divide. In A. A. Summers, P. C. Cheshire, & L. Senn (Eds.), *Urban change in the United States and Western Europe: Comparative analysis and policy* (pp. 581-592). Washington, DC: Urban Institute.

Kling, R., Olin, S., & Poster, M. (Eds.). (1991). *Postsuburban California: The transformation of Orange County since World War II*. Berkeley: University of California Press.

Knaap, G., & Nelson, A. C. (1992). *The regulated landscape: Lessons on state land use planning from Oregon*. Cambridge, MA: Lincoln Institute of Land Policy.

Knights, D., & Morgan, G. (1990). The concept of strategy in sociology. *Sociology, 24*(3), 475-483.

Koff, S. (1996, January 9). Council balks on Gateway bail-out. *Cleveland Plain Dealer,* p. B-1.

Koff, S., Heider, T., & Grossi, T. (1996, February 10). NFL owners OK deal. *Cleveland Plain Dealer,* p. A-1.

Kotz, D. M., McDonough, T., & Reich, M. (Eds.). (1994). *Social structures of accumulation: The political economy of growth and crisis*. Cambridge, UK: Cambridge University Press.

Krätke, S. (1995). *Stadt, Raum, Oekonomie: Einfuehrung in aktuelle Problemfelder der Stadtoekonomie und Wirtschaftsgeographie*. Basel, Switzerland: Birkhaueser Verlag.

Krätke, S., & Schmoll, F. (1991). The local state and social restructuring. *International Journal of Urban and Regional Research, 15*(1), 542-552.

Lauria, M. (1986). Toward a specification of the local state: State intervention strategies in response to a manufacturing plant closure. *Antipode, 18*(1), 39-65.

Lauria, M. (1994a). The transformation of local politics: Manufacturing plant closures and governing coalition fragmentation. *Political Geography, 13*(6), 515-539.

Lauria, M. (1994b, July). *Waterfront development, urban regeneration and local politics in New Orleans and Liverpool*. Paper presented at the conference Shaping the Urban Future: International Perspectives and Exchanges, Bristol, University of Bristol, School for Advanced Urban Studies.

Lauria, M., & Whelan, R. K. (1995). Planning theory and political economy: The need for reintegration. *Planning Theory, 14*, 8-33.

Lawless, P. (1994). Partnership in urban regeneration in the UK: The Sheffield central area study. *Urban Studies, 31*(8), 1303-1324.

Leitner, H. (1990). Cities in pursuit of economic growth: The local state as entrepreneur. *Political Geography Quarterly, 9*, 146-170.

Leitner, H., & Garner, M. (1993). The limits of local initiatives: A reassessment of urban entrepreneurialism for urban development. *Urban Geography, 14*, 57-77.

Leo, C. (1977). *The politics of urban development: Canadian urban expressway disputes*. Toronto: Institute of Public Administration of Canada.

Leo, C. (1994). The urban economy and the power of the local state: The politics of planning in Edmonton and Vancouver. In F. Frisken (Ed.), *The changing Canadian metropolis: A public policy perspective* (Vol. 2). Berkeley: Institute of Governmental Studies Press, University of California.

Leo, C. (1995a). Global change and local politics: Economic decline and the local regime in Edmonton. *Journal of Urban Affairs, 17*(3), 277-299.

Leo, C. (1995b). The state in the city: A political economy perspective on growth and decay. In J. Lightbody (Ed.), *Canadian metropolitics: Governing our cities* (pp. 27-50). Toronto: Copp Clark.

Leo, C., & Fenton, R. (1990). Mediated enforcement and the evolution of the state: Urban development corporations in Canadian city centres. *International Journal of Urban and Regional Research, 14*(2), 185-206.

Ley, D. (1994). Social polarization and community response: Contesting marginality in Vancouver's Downtown Eastside. In F. Frisken (Ed.), *The changing Canadian metropolis: A public policy perspective* (Vol. 2). Berkeley: Institute of Governmental Studies Press, University of California.

Lindblom, C. E. (1977). *Politics and markets: The world's political-economic systems*. New York: Basic Books.

Lipietz, A. (1986). Behind the crisis: The exhaustion of a regime of accumulation. A regulation school perspective on some French empirical works. *Review of Radical Political Economics, 18,* 13-32.

Lipietz, A. (1987). *Mirages and miracles: The crisis of global fordism* (D. Macey, Trans.). London: Verso.

Lipietz, A. (1988, Summer). Reflections on a tale: The Marxist foundations of the concept of regulation and accumulation. *Studies in Political Economy, 26,* 7-36.

Lipietz, A. (1992). *Towards a new economic order.* Cambridge, UK: Polity.

Lipietz, A. (1993). The local and the global: Regional individuality or interregionalism? *Transactions of the Institute of British Geographers, 18*(1), 8-18.

Lipietz, A. (1994). The national and the regional: Their autonomy vis-à-vis the capitalist world crisis. In R. P. Palan & B. Gills (Eds.), *Transcending the state-global divide: A neo-structuralist agenda in international relations* (pp. 23-44). Boulder, CO: Lynne Riener.

Lipset, S. M. (1990). *Continental divide: The values and institutions of the United States and Canada.* New York: Routledge.

Logan, J., & Molotch, H. (1987). *Urban fortunes: The political economy of place.* Berkeley: University of California Press.

Logan, J., & Swanstrom, T. (Eds.). (1990). *Beyond the city limits.* Philadelphia: Temple University Press.

Longhurst, B. (1991). Raymond Williams and local cultures. *Environment and Planning A, 23,* 229-238.

Lordon, F. (1995, November). *L'etat et l'economie politique.* Paper presented to Colloque de Politique Economique, Paris.

Lorimer, J. (1978). *The developers.* Toronto: Lorimer.

Lorenz, E. H. (1993). Flexible production systems and the social construction of trust. *Politics and Society, 21,* 307-324.

Logue, E. Papers. Manuscripts and archives, Stirling Memorial Library, Yale University, New Haven, CT. Boxes 153, 217.

Lovering, J. (1990). Fordism's unknown successor: A comment on Scott's theory of flexible accumulation and the re-emergence of regional economies. *International Journal of Urban and Regional Research, 14,* 159-174.

Lowe, J. R. (1967). *Cities in a race with time.* New York: Random House.

Magnet, M. (1995). How business bosses saved a sick city. In W. D. Keating, N. Krumholz, & D. C. Perry (Eds.), *Cleveland: A metropolitan reader* (pp. 155-160). Kent, OH: Kent State University Press.

Malecki, E. J., & Bradbury, S. L. (1992). R&D facilities and professional labour: Labour force dynamics in high technology. *Regional Studies, 26*(2), 123-136.

Mann, M. (1984). The autonomous power of the state: Its origins, mechanisms and results. In M. Mann (Ed.), *States, war and capitalism* (pp. 1-33). Oxford, UK: Blackwell.

Mann, M. (1988). The autonomous power of the state: Its origins, mechanisms, and results. In M. Mann (Ed.), *States, war and capitalism* (pp. 1-32). Oxford, UK: Blackwell.

Mann, M. (1993). *The sources of social power: Vol. 2. The rise of classes and nation-states, 1760-1919.* Cambridge, UK: Cambridge University Press.

Marchione, W. (1976, September). The 1949 charter reform. *New England Quarterly,* pp. 373-398.

Marglin, S. A. (1974). What do bosses do? *Review of Radical Political Economics, 6,* 60-92.

Markusen, A. R., & Bloch, R. (1985). Defensive cities: Military spending, high technology, and human settlement. In M. Castells (Ed.), *High technology, space, and society* (pp. 106-120). Beverly Hills, CA: Sage.

Massey, D. (1994). *Time, place and gender.* Cambridge, UK: Polity.

Mayer, M. (1994). Post-fordist city politics. In A. Amin (Ed.), *Post-fordism: A reader* (pp. 316-337). Oxford, UK: Blackwell.

Mensch, G. O. (1975). *Stalemate in technology: Innovations overcome the depression.* Cambridge, MA: Ballinger.

Mier, R. (Ed.). (1993). *Social justice and local development policy.* Newbury Park, CA: Sage.

Miggins, E. M. (1995). Between spires and stacks: The people and neighborhoods of Cleveland. In W. D. Keating, N. Krumholz, & D. C. Perry (Eds.), *Cleveland: A metropolitan reader* (pp. 179-201). Kent, OH: Kent State University Press.

Miller, C. P., & Wheeler, R. A. (1995). Cleveland: The making and remaking of an American city. In W. D. Keating, N. Krumholz, & D. C. Perry (Eds.), *Cleveland: A metropolitan reader* (pp. 31-48). Kent, OH: Kent State University Press.

Miller, G. (1981). *Cities by contract: The politics of municipal incorporation.* Cambridge: MIT Press.

Miron, L. (1992). Corporate ideology and the politics of entrepreneurism in New Orleans. *Antipode, 24*(4), 263-288.

Mollenkopf, J. H. (1983). *The contested city.* Princeton, NJ: Princeton University Press.

Molotch, H. L. (1976). The city as a growth machine: Towards a political economy of place. *American Journal of Sociology, 82,* 309-331.

Molotch, H., & Vicari, S. (1988). Three ways to build: The development process in the United States, Japan, and Italy. *Urban Affairs Quarterly, 24*(2), 188-210.

Morishima, M. (1973). *Marx's economics: A dual theory of value and growth.* Cambridge, UK: Cambridge University Press.

Nelson, A. C. (1995). Growth management and the savings-and-loan bailout. *The Urban Lawyer, 27*(1), 71-85.

Newman, P. C. (1982). *The acquisitors.* Toronto: Seal.

Newton, K. (1975). American urban politics: Social class, political structure and public goods. *Urban Affairs Quarterly, 11*(2), 241-264.

Nisbet, A. (1991). *The textile industry in New England: Is flexible specialization occurring?* Masters Research Project, University of Rhode Island Graduate Curriculum in Community Planning and Area Development.

Noël, A. (1988). *Action collective, partis politiques et relations industrielles: Une logique politique pour l'approche de la regulation.* International Conference on Regulation Theory, Barcelona.

Noyelle, T., & Stanback, T. M., Jr. (1984). *The economic transformation of American cities.* Totowa, NJ: Rowman & Allanheld.

O'Connor, T. H. (1993). *Building a new Boston, politics and urban renewal, 1950-1970.* Boston: Northeastern University.

Olson, M. (1965). *The logic of collective action.* Cambridge, MA: Harvard University Press.

Orr, M. E., & Stoker, G. (1994). Urban regimes and leadership in Detroit. *Urban Affairs Quarterly, 30*(1), 48-73.

Ozanian, M. K. (1995, May 9). Suite deals: Why new stadiums are shaking up the pecking order of sports franchises. *Financial World, 164*(11), 42-56.

Pagano, M. A., & Bowman, A. O. (1995). *Cityscapes and capital.* Baltimore: Johns Hopkins University Press.

Painter, J. (1991). Regulation theory and local government. *Local Government Studies, 17*(6), 23-44.

Painter, J. (1995). Regulation theory, post-fordism and urban politics. In D. Judge, G. Stoker, & H. Wolman (Eds.), *Theories of urban politics* (pp. 276-295). London: Sage.

Painter, J., & Goodwin, M. (1995). Local governance and concrete research: Investigating the uneven development of regulation. *Economy and Society, 24*(3), 334-356.

Painter, J., Wood, M., & Goodwin, M. (1995, September). *British local governance after fordism: A regulationist interpretation.* Paper presented at the ESRC Local Governance Research Programme conference, Exeter, UK.

Peck, J., & Tickell, A. (1992). Local modes of social regulation? Regulation theory, Thatcherism and uneven development. *Geoforum, 23,* 347-364.

Peck, J., & Tickell, A. (1994). Searching for a new institutional fix: The after-fordist crisis and the global-local disorder. In A. Amin (Ed.), *Post-fordism: A reader* (pp. 280-315). Oxford, UK: Blackwell.

Perkins, G. (1990). Boston gross city product and productivity: 1972-1986 with geographic comparisons. *Technical appendix to the Howell Report* (Vol. 1). Boston: Boston Redevelopment Authority.

Peterson, I. (1995, September 10). The "mistake" wakes up, roaring. *New York Times,* p. A-9.

Peterson, P. (1981). *City limits.* Chicago: University of Chicago Press.

Petshek, K. R. (1973). *The challenge of urban reform.* Philadelphia: Temple University Press.

Philadelphia City Planning Commission. (1954, September). Planning in Philadelphia. *ASPO Newsletter, 20*(9), 76-80.

Pickvance, C., & Preteceille, E. (1991). *State restructuring and local power: A comparative perspective.* London: Pinter.

Pincetl, S. (1994). The regional management of growth in California: A history of failure. *International Journal of Urban and Regional Research, 18,* 256-274.

Piore, M. J., & Sabel, C. F. (1984). *The second industrial divide: Possibilities for prosperity.* New York: Basic Books.

Piven, F. F., & Friedland, R. (1984). Public choice and private power: A theory of the urban fiscal crisis. In A. Kirby, P. Knox, & S. Pinch (Eds.), *Public service provision and urban development* (pp. 390-420). New York: Croom Helm/St. Martin's.

Plotkin, S. (1987). *Keep out: The struggle for land use control.* Berkeley: University of California Press.

Plotkin, S. (1990). Enclave consciousness and neighborhood activism. In J. Kling & P. Posner (Eds.), *Dilemmas of community activism* (pp. 218-239). Philadelphia: Temple.

Polanyi, K. (1944). *The great transformation.* New York: Farrar & Rinehart.

Poulantzas, N. (1978). *State, power, socialism.* London: New Left Books.

Powers, J. (1986, June 1). On the move with Steve Coyle. *Boston Globe Magazine,* pp. 23-40.

Ragonetti, T. (1977, April 22). *Informal tax agreements in Boston: Analysis, evaluation and directions for change, a report to Collector—Treasurer James V. Young.*

Real Estate Research Corporation. (1974). *Urban renewal land disposition study.* Chicago.

Rosentraub, M. S., Swindell, D., Przybylski, M., & Mullins, D. R. (1994). Sport and downtown development strategy: If you build it, will jobs come? *Journal of Urban Affairs, 16*(3), 221-239.

Rutchick, J., Heider, T., & Koff, S. (1995, July 9). How Gateway went awry. *Cleveland Plain Dealer,* p. A-1

Rutchick, J., & Koff, S. (1995, June 27). Chema quits as Gateway director. *Cleveland Plain Dealer,* p. A-1.

Rycroft, C. (1990). The internationalization of U.S. intergovernmental relations in science and technology policy. *Technology in Society, 12,* 217-233.

San Bernardino County. n.d. *Memorandum of understanding by and between the U.S. Fish and Wildlife Service, the California Department of Fish and Game, the County of San Bernardino, the fifteen affected cities in southwestern San Bernardino County and additional undersigned participating agencies for the purpose of developing and implementing a habitat conservation plan to conserve wildlife and plant species of concern in the San Bernardino Valley.* Author draft mimeo.

San Bernardino/Riverside Counties. (1994). *San Bernardino/Riverside Counties economic recovery/defense conversion strategic plan* (Report No. 8), San Bernardino, CA.

Salamon Brothers. (1989, December, 13). The Boston office market: Will it survive the Massachusetts miracle? New York: Author.

Sancton, A. (1991). The municipal role in the governance of Canadian cities. In T. Bunting & P. Filion (Eds.), *Canadian cities in transition.* Toronto: Oxford University Press.

Sanders, H. T., & Stone, C. N. (1987). Developmental politics reconsidered. *Urban Affairs Quarterly, 22*(4), 521-539.

Sandomir, R. (1996, February 12). Compromise got Cleveland a stadium. *New York Times,* p. B-9.

Sassen, S. (1991). *The global city: New York, London, Tokyo.* Princeton, NJ: Princeton University.

Savitch, H. V. (1988). *Post-industrial cities: Politics and planning in New York, Paris and London.* Princeton, NJ: Princeton University Press.

Savitch, H. V., & Kantor, P. (1994, July). *City business: An international perspective on market place politics.* Paper presented at the conference Shaping the Urban Future: International Perspectives and Exchanges, Bristol, University of Bristol, School for Advanced Urban Studies.

Saxenian, A. (1994). *Regional advantage: Culture and competition in Silicon Valley and Route 128.* Cambridge, MA: Harvard University Press.

Sayer, A. (1984). *Method in social science: A realist approach* (1st ed.). London: Hutchison.

Sayer, A. (1989). Post-fordism in question. *International Journal of Urban and Regional Research, 13,* 666-693.

Sayer, A. (1992). *Method in social science: A realist approach* (2nd ed.). London: Routledge Kegan Paul.

Sayer, A., & Walker, R. (1992). *The new social economy: Reworking the division of labor.* Cambridge, UK: Blackwell.

Schattschneider, E. E. (1970). *The semi-sovereign people.* Hinsdale, IL: Dryden.

Schumpeter, J. S. (1939). *Business cycles.* New York: McGraw-Hill.

Scott, A. J. (1986). Industrial organization and location: Division of labor, the firm, and spatial process. *Economic Geography, 62,* 215-231.

Scott, A. J. (1988). *New industrial spaces: Flexible production organization and regional development in North America and Europe.* Berkeley: University of California Press.

Scott, A. J. (1990). *Metropolis: From the division of labor to urban form.* Berkeley: University of California.

Scott, A. J., & Paul, A. (1990). Collective order and economic coordination in industrial agglomerations: The technopoles of Southern California. *Environment and Planning C: Government and Policy, 8,* 179-193.

Schultz, E., & Simmons, W. (1959). *Offices in the sky.* Indianapolis: Bobbs-Merrill.

Sege, I. (1992, May 28). Mass. residents shared 1980s boom unequally. *Boston Globe,* p. 1.

Sewell, J. (1993). *The shape of the city: Toronto struggles with modern planning.* Toronto: University of Toronto Press.

Shatten, R. A. (1995). Cleveland Tomorrow: A practicing model of new roles and processes for corporate leadership in cities. In W. D. Keating, N. Krumholz, & D. C. Perry (Eds.), *Cleveland: A metropolitan reader* (pp. 321-331). Kent, OH: Kent State University Press.

Shefter, M. (1985). *Political crisis/fiscal crisis: The collapse and revival of New York City.* New York: Basic Books.

Sheppard, E., & Barnes, T. J. (1990). *The capitalist space economy: Geographical analysis after Ricardo, Marx and Sraffa.* London: Unwin Hyman.

Shropshire, K. L. (1995). *The sports franchise game: Cities in pursuit of sports franchises, stadiums, and arenas.* Philadelphia, PA: University of Pennsylvania Press.

Slavet, J. (1977, November). Tax policy analysis and planning study, the Boston case (HUD Research Grant No. 2628). *Second Project Monitor.*

Smith, M. P., & Feagin, J. R. (1987). *The capitalist city: Global restructuring and community politics.* Oxford, UK: Blackwell.

Snyder, S. (1988, December 27). Where have all the powers gone? *Boston Globe,* p. 33.

Soja, E. W. (1989). *Postmodern geographies: The reassertion of space in critical social theory.* London: Verso.

Sorkin, M. (Ed.). (1992). *Variations on a theme park: The new American city and the end of public space*. New York: Noonday.

Stoker, G. (1990). Regulation theory, local government and the transition from fordism. In D. S. King & J. Pierre, *Challenges to local government* (pp. 242-264). London: Sage.

Stoker, G. (1995). Regime theory and urban politics. In D. Judge, G. Stoker, & H. Wolman (Eds.), *Theories of urban politics* (pp. 54-71). London: Sage.

Stoker, G., & Mossberger, K. (1994). Urban regime theory in comparative perspective. *Environment and Planning C: Government and Policy, 12*, 195-212.

Stone, C. N. (1987). The study of the politics of urban development. In C. N. Stone & H. T. Sanders (Eds.), *The politics of urban development* (pp. 3-22). Lawrence: University Press of Kansas.

Stone, C. N. (1989). *Regime politics: Governing Atlanta, 1946-1988*. Lawrence: University of Kansas Press.

Stone, C. N. (1991). The hedgehog, the fox, and the new urban politics: A rejoinder to Kevin R. Cox. *Journal of Urban Affairs, 13*(3), 289-297.

Stone, C. N. (1993). Urban regimes and the capacity to govern: A political economy approach. *Journal of Urban Affairs, 15*, 1-28.

Stone, C. N., Orr, M. E., & Imbroscio, D. (1991). The reshaping of urban leadership in U.S. cities: A regime analysis. In M. Gottdiener & C. G. Pickvance (Eds.), *Urban life in transition* (Vol. 39, pp. 222-239). Newbury Park, CA: Sage.

Stone, C. N., & Sanders, H. T. (Eds.). (1987). *The politics of urban development*. Lawrence: University Press of Kansas.

Storper, M. (1991, April 13-17). *Regional "worlds of production": Conventions of learning and innovation in flexible production systems of France, Italy, and the USA*. Presented at the annual meeting of the Association of American Geographers, Miami.

Storper, M., & Harrison, B. (1990). *Flexibility, hierarchy, and regional development: The changing structure of industrial production systems and their forms of governance in the 1990s*. Graduate School of Architecture and Urban Planning. Los Angeles: University of California Press.

Storper, M., & Walker, R. (1989). *The capitalist imperative: Territory, technology and industrial growth*. Oxford, UK: Basil Blackwell.

Sum, N-L. (1995, November). *Politics of identity: A temporal-spatial perspective on Greater China*. Paper presented at the East Asia Research Centre, School of East Asian Studies, University of Sheffield.

Swanstrom, T. (1985). *The crisis of growth politics: Cleveland, Kucinich, and the challenge of urban populism*. Philadelphia: Temple University Press.

Swanstrom, T. (1995). Urban populism, fiscal crisis, and the new political economy. In W. D. Keating, N. Krumholz, & D. C. Perry (Eds.), *Cleveland: A metropolitan reader* (pp. 97-118). Kent, OH: Kent State University Press.

Sweeney, S. M. (1994, December). Where's Roldo? *Northern Ohio Live*, pp. 19-21, 45.

Swyngedouw, E. A. (1991, April). *"Spatial organization" as a force of production and the space/technology nexus*. Presented at the annual meeting of the Association of American Geographers, Miami.

Thrift, N. (1983). On the determination of social action in space and time. *Environment and Planning D: Society and Space, 1*(1), 23-57.

Thrift, N. (1989, September). *Flexible production in financial services. Is it new? Does it matter?* Presented at the Cardiff Symposium on Regulation, Innovation, and Spatial Development, Cardiff, Wales, UK.

Tickell, A., & Peck, J. A. (1992). Accumulation, regulation and the geographies of post-fordism: Missing links in regulationist research. *Progress in Human Geography, 16*(2), 190-218.

Tickell, A., & Peck, J. A. (1995). Social regulation after fordism: Regulation theory, neo-liberalism and the global-local nexus. *Economy and Society, 24*, 357-386.

Tinkcom, M. B. (1982). Depression and War: 1929-1946. In R. F. Weigley (Ed.), *Philadelphia: A 300-year history* (pp. 601-648). New York: Norton.

U.S. Bureau of the Census. (1951, 1962, 1970, 1980, 1990). *County business patterns.* Washington, DC: Government Printing Office.

Van Allsburg, C. M. (1986). Dual-earner housing needs. *Journal of Planning Literature, 1,* 388-399.

Van der Pijl, K. (1982). *The making of the Atlantic ruling class.* London: Verso.

Vicari, S., & Molotch, H. (1990). Building Milan: Alternative machines of growth. *International Journal of Urban and Regional Research, 14*(4), 602-624.

Vickers, R. J. (1996, February 6). Seating licenses seen as option. *Cleveland Plain Dealer,* p. A-1.

Vigman, F. K. (1955). *Crisis of the cities.* Washington, DC: Public Affairs Press.

Walker, D. (1979). *The great Winnipeg dream: The redevelopment of Portage and Main.* Oakville, Ontario: Mosaic Press.

Walker, R. A. (1988). The dynamics of value, price and profit. *Capital & Class, 35,* 146-181.

Walker, R. A. (1989). A requiem for corporate geography: New directions in industrial organization, the production of place and uneven development. *Geografiska Annaler, 71*(B), 43-68.

Walker, R. A. (1995). Regulation and flexible specialization as theories of capitalist development: Challengers to Marx and Schumpeter? In H. Liggett & D. C. Perry (Eds.), *Spatial practices: Critical explorations in social/spatial theory* (pp. 167-208). Thousand Oaks, CA: Sage.

Warner, K., & Molotch, H. L. (1995). Power to build: How development persists despite local controls. *Urban Affairs Review, 30,* 378-406.

Watson, W. (1990). Strategy, rationality and inference: The possibility of symbolic performances. *Sociology, 24*(3), 485-498.

Weinberg, M. (1981). Boston's Kevin White: A mayor who survives. *Political Science Quarterly, 96*(1), 87-106.

Whelan, R. K. (1987). New Orleans: Mayoral politics and economic-development policies in the postwar years 1945-86. In C. N. Stone & H. T. Sanders (Eds.), *The politics of urban development* (pp. 216-222). Lawrence: University Press of Kansas.

Whelan, R. K., Young, A. H., & Lauria, M. (1994). Urban regimes and racial politics in New Orleans. *Journal of Urban Affairs, 16*(1), 1-21.

Whitman, D., & Friedman, D. (1994, October 17). The white underclass. *U.S. News and World Report,* 40-53.

Williams, R. (1961). *The long revolution.* Harmondsworth, UK: Penguin.

Williamson, O. (1975). *Markets and hierarchies.* New York: Free Press.

Wilson, D. (1995). Representing the city: Growth coalitions and uneven development in two midwest cities. *Planning Theory, 14,* 96-115.

Wolf, S. G. (1982). The bicentennial city: 1968-1982. In R. F. Weigley (Ed.), *Philadelphia: A 300-year history* (pp. 704-734). New York: Norton.

Wolfe, A. (1981). *America's impasse.* New York: Pantheon.

Wood, A. M. (1993). Local economic development networks and prospecting for industry. *Environment and Planning A, 25,* 1649-1662.

Woolley, S. (1992, June 5). Inland Empire steps up effort to lure business from neighboring Los Angeles. *Wall Street Journal,* p. A2.

Yin, R. (1980). Creeping federalism: The federal impact on the structure and function of local government. In N. Glickman (Ed.), *The urban impact of federal policies* (pp. 595-618). Baltimore: Johns Hopkins University Press.

Zimbalist, A. (1992). *Baseball billions: A probing look inside the big business of our national pastime.* New York: Basic Books.

# Name Index

# Subject Index

262

# About the Contributors

**Robert A. Beauregard** is Professor in the Milano Graduate School of Management and Urban Policy at the New School for Social Research, where he teaches urban political economy and economic development. He has a Ph.D. in city and regional planning from Cornell University and previously taught at the University of Pittsburgh and Rutgers University. He was Harvey Perloff Visiting Professor in the Graduate School of Architecture and Urban Planning at the University of California, Los Angeles. His most recent book is *Voices of Decline: The Postwar Fate of U.S. Cities.*

**Kevin R. Cox** is Professor of Geography at Ohio State University and Distinguished Visiting Professor at Reading University. His research interests include the politics of local economic development, critical human geography, and the political economy of South Africa. He is the author of *Location and Public Problems and Conflict* and *Power and Politics in the City,* editor of *Urbanization and Conflict in Market Societies,* and coeditor of *Conflict, Politics, and the Urban Scene and Behavioral Geography Revisited.* His papers have appeared in numerous journals, including *Acta Sociologica, Urban Studies,* the *Journal of Urban Affairs,* the *International Journal of Urban and Regional Research, Society and Space, Urban Geography, Political Geography,* and *Geografiska Annaler.*

**Marshall M. A. Feldman** is Associate Professor of Community Planning at the University of Rhode Island. He also has taught at Cleveland State University; the University of California, Berkeley; San Francisco State University; the University of California, Santa Cruz; and the University of Texas. He holds a Ph.D. in urban planning from the University of California, Los Angeles, and received a B.S. and a Masters of Engineering (Industrial) from Cornell University. His research is on housing and economic development and more generally on urbanization and urban political economy. He has written in these areas and on planning education, planning theory, and methodology.

**Mark Goodwin** is Professor of Human Geography in the Institute of Earth Studies at the University of Wales, Aberystwyth. He received his doctorate from the London School of Economics. He has published widely in the fields of urban politics, urban geography, and regional change. He is the joint author of two books on the local state and local politics, *Housing States and Localities* and *The Local State and Uneven Development.* He has two books forthcoming on urban change. He is also a joint author of *Practising Human Geography,* which explores the construction and interpretation of geographical data. He has directed several research projects; the latest, with Joe Painter, examines changes in the structures and practices of local governance.

**Cynthia Horan** is Visiting Assistant Professor of African American Studies and Government at Wesleyan University in Connecticut. She previously taught at the University of Toronto. She earned a Ph.D. in urban studies and planning from the Massachusetts Institute of Technology and has published articles in the *Journal of Urban Affairs, Polity,* and *Research in Political Economy.* She is currently writing a book on the politics of Boston's postwar economic transformation. Her research interests include urban political economy, state theory, and racial politics.

**Bob Jessop** is Professor of Sociology at Lancaster University, England. Previously, he taught in the Department of Government at Essex University. He has written extensively on theories of the state, the political economy of Thatcherism, the regulation approach, postsocialist economies, and social theory. He is currently completing a research project on the transformation of local governance in Britain and continuing work on regionalism in East Asia. His books include *The Capitalist State, Nicos Poulantzas, Thatcherism: A Tale of Two Nations, State Theory,* and *The Politics of Flexibility.* His work on the regulation approach and governance has appeared in *Economy and Society* and various edited collections.

**Andrew E. G. Jonas** is Lecturer in the School of Geography and Earth Resources, University of Hull, United Kingdom, and Adjunct Assistant Professor in the Department of Earth Sciences, University of California, Riverside. His research interests are U.S. urban policy and politics, labor and community responses to deindustrialization, and conservation policy in California. He has published articles in *Economic Geography, Society and Space, Journal of Urban Affairs, Transactions of the Institute of British Geographers, Urban Geography, Political Geography, Antipode,* and *Area.* He currently has a 2-year NSF project on habitat conservation planning and urban development in Southern California and is coediting a book on critical perspectives on the city as a growth machine.

**W. Dennis Keating** is Professor of Urban Planning and Law, and Associate Dean of the Levin College of Urban Affairs at Cleveland State University. He has a Ph.D. in City and Regional Planning from the University of California, Berkeley, and a J.D. from the University of Pennsylvania Law School. His latest books are *Revitalizing Urban Neighborhoods, Cleveland: A Metropolitan Reader,* and *The Suburban Racial Dilemma.* He is currently editing books on distressed central city areas and rent control.

**Mickey Lauria** is Professor of Urban and Regional Planning and Director of the Division of Urban Research and Policy Studies in the College of Urban and Public Affairs at the University of New Orleans. His earned his A.B. in political science and geography from the University of California, Los Angeles, and his M.A. and Ph.D. in geography from the University of Minnesota. He is coeditor of the *Journal of Planning Education and Research.* He has published articles on urban redevelopment, urban politics, and community-based development organizations in planning, geography, and urban studies journals. His recent local research interests also include patterns and impacts of housing foreclosures, historical analysis of preservation conflicts, and planning issues involving race and class in New Orleans.

**Christopher Leo** is Professor of Political Science and Coordinator of Urban Studies at the University of Winnipeg. He holds a Ph.D. in Political Economy from the University of Toronto. Recent articles have appeared in the *International Journal of Urban and Regional Research, Journal of Urban Affairs, The Changing Canadian Metropolis: Contemporary Perspectives,* and *Canadian Metropolitics.* Recent and current research deals with the politics of development in Vancouver, Edmonton, Winnipeg, and Toronto;

the politics of growth management in Portland, Oregon; comparison of European and North American urban politics; and the impact of global change on local politics.

**Joe Painter** is Lecturer in Geography at the University of Durham. He received his B.A. from Cambridge University and his Ph.D. from the Open University. He was previously Lecturer in Geography at the University of Wales, Lampeter. He is the author of several articles on regulation theory, urban politics, and the restructuring of the local state and of the book *Politics, Geography, and "Political Geography."* He is currently completing a major research project on the changing role of the British local state.